DNA SIMPLIFIED:

The Hitchhiker's Guide to DNA

Everything You Always Wanted To Know About DNA
(so you could sound really intelligent
at cocktail parties and staff meetings)

Daniel H. Farkas, Ph.D.

AACC Press
2101 L Street, N.W.
Washington, D.C. 20037-1526

Cover graphics by Paul Thiessen.

Front cover: CPK with stone
Inside front cover: Outdoors
Inside back cover: Model on a stand

Paul Thiessen is currently a Ph.D. student at the University of Illinois at Urbana-Champaign in the organic chemistry department, and eventually hopes to have a career as a computational biomolecular chemist. Paul enjoys computer graphics and programming as a hobby, especially when combining science with art.

Cover design by Gaye Roth Watts.

ISBN 0-915274-84-1

Printed in the United States of America

I'd like to dedicate this book to those who are most important in my life: my wife, Becky, and my children, Joshua and Haley; my parents (I miss you Dad), my brothers and their families, and my close friends and the rest of my family. I'd also like to acknowledge my co-workers, Fritz, Domnita, and Deanna, who keep it all interesting. Special thanks go out to Steve, who planted the seed for all of this.

Preface

Everyone seems to be interested in DNA. It's no wonder; not only do we and every living thing all have DNA but we're constantly being bombarded with information about DNA. DNA is on the TV and radio news, it's in the papers, Nobel Laureate Kary Mullis is even making sure that it's going to be available as jewelry. There are advertising agencies, marketing agencies, fragrances and board games named after DNA, and there are countless more examples. DNA is something of a cultural icon, as are the associated topics of genetics and heredity, and will no doubt influence our culture and society greatly as we move into the 21st century. As a global society, we will need to deal with the vast implications of research into our genetic makeup. These implications include but are not limited to medicine, privacy, insurability, ethics, patents and other business issues, matters of paternity and immigration, criminal investigations, and probably extend to military applications.

Aside from the cultural phenomenon over DNA that we are witnessing, DNA has been an intense subject of scientific and clinical investigation for many years now. The basic structure of DNA, the famed double-helix, was worked upon throughout the late 1940s and early 1950s and finally deduced and published by James Watson and Francis Crick in April 1953. This is a good point in time to designate as the birth of molecular pathology. Molecular pathology is a relatively new clinical laboratory discipline that really took off with the technological advances of the Southern blot and the polymerase chain reaction (PCR), both of which are described in the text of this book. As our insight into human disease deepens and our understanding of the role of DNA and heredity in the pathogenesis of disease increases, molecular pathology continues to take on an increasingly important clinical role.

With that importance comes appropriate concern over issues of privacy, confidentiality, ethics, insurability, etc.; in other words "really heavy stuff." I like to think that our society, driven as it is by markets and politics, will adequately address these important issues. At the same time, as deeply involved as I have been in molecular pathology for the last fifteen years or so, I realize that everyone is interested in DNA. When a taxi driver is taking me from the airport to the hotel that I'll be using while I attend a scientific meeting, I invariably get asked what I do. A spirited conversation usually follows which is punctuated by questions of genuine interest and curiosity (often, "When are you guys gonna cure cancer already?" or "So do you think this AIDS thing was a government plot?"). My friends and family want to understand DNA and its implications for health; the receptionist at the dentist's office prattles on about DNA ("Gee, some day it might even be important in dental care." "You're right," I tell her.) There's a thirst out there for more information that is not peppered with the technobabble by which so many people get turned off and so many scientists and physicians use to guard their professional stations.

If you think that DNA stands for "don't [k]now anything" then I hope you will find this book both useful and perhaps a bit entertaining at the same time. The idea came to my wife and me while working on one of my other books that she happened to be indexing at the time. There seemed to be a lot of similarity between two biochemical reactions that purported themselves as "PCR alternatives." When we sat down to work out the biochemistry of how the two reactions actually differed, it became clear that for all intents and purposes, one was the other with a trademark symbol. So we came to realize that a lot of this stuff, which we had learned had some rather wide general appeal, was overly complicated and could be simplified to the point where it was easy to understand.

DNA Simplified: The Hitchhiker's Guide to DNA is meant to be an authoritative, factually correct, yet somewhat lighthearted look at the practice of clinical molecular pathology and the associated topics of DNA and genetics. I have intentionally used a casual writing style because I find it more enjoyable to write that way and I think it will help make the book read more easily and that its contents will therefore be easier to understand. I have organized the book in the manner of *The Hitchhiker's Guide to Clinical Chemistry* (also published by AACC Press), i.e., I have followed the same outline of alphabetical listing of subjects and items. The book has entries for most of the common terms used in molecular pathology and DNA technology, but I have strived not to be esoteric. The entries have meaning to any scientist or physician and, in fact, to any educated (or eager to learn) individual. Indeed, I do not mean for the book to be restricted to use by professionals. I like to think that strengths of the book will include its brevity, its language to promote ease of understanding, and its general appeal since so many of us seem to be interested in DNA. Please enjoy it and share it with your friends.

Daniel H. Farkas, Ph.D.
Rochester Hills, Michigan

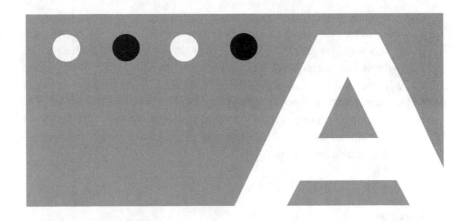

There are quite a few references throughout to laboratory procedure and protocol. You may come to appreciate that there are a lot of laboratory procedures that are a lot like following a recipe. In general, I have found it to be true that if you are good in the kitchen, you're good in the laboratory and vice versa.

-ase

A suffix which denotes that something is an enzyme: a protein that has a specific biochemical job to do. You'll see words ending in "ase" throughout this book. [☞ Nuclease (DNase and RNase)]

A-DNA

Companies like to call themselves AAA, Inc. or AAAA Widgets, Inc., so they can be listed first in the phone book. Well, why would you think DNA is any different?

DNA naturally forms as a long, double-helix shaped, stringlike structure inside cells. DNA is made up of two strands wound around each other in a right-handed coil (apologies to all you southpaws). The strands are made up of chemical compounds called nucleotides; these are ring-shaped structures composed of nitrogen, oxygen, phosphorus,

1

carbon, and hydrogen (so don't forget to take your vitamin pill every day). The nucleotides bind to each other on opposite strands of the helix in a defined way. [☞ also Complementary Strands of DNA] The natural way in which the nucleotides bind generates the form of DNA that is ordinarily found in living cells, called the B form of DNA, or B-DNA. [☞ B-DNA] Under unusual laboratory conditions, the way the bases bind to each other can be changed subtly so that unusual shape, angles of binding, and distances in the DNA molecule occur within the DNA double helix; that's what's known as A-DNA.

Agarose

Similar to Jell-O®, but doesn't taste as good (actually, we never taste the stuff in lab; OSHA, the Federal Occupational Safety and Health Administration, would throw us all in jail). The consistency of agarose gels is somewhere between that of Jell-O Jigglers and a Jell-O mold. We mix powdered agarose (derived from seaweed) with water and some salts and microwave it until it boils. We wait for it to cool and then pour the liquefied agarose solution into our Jell-O mold, which is an electrophoresis chamber. We then place a comb with teeth into the liquefied agarose, and the agarose hardens around the teeth. After a short time, when the agarose is hard to the touch, we remove the comb. Indentations or wells in the agarose have been formed, and we add our DNA solution to these wells so it can be analyzed by electrophoresis. [☞ also Electrophoresis]

Allele

All cells in the human body are diploid, which means that they have two full sets of DNA-containing chromosomes. Two sets of 23 comprise the 46 chromosomes found in human cells. There are exceptions. There is no DNA in mature red blood cells. Human sex cells, also called gametes, which are the sperm in males and the eggs in females, have only one set of chromosomes. Sometimes tumor cells, being the ornery, unpredictable, unwelcome critters that they are, don't have the normal complement of chromosomes; they can be diploid, triploid (three sets of chromosomes), or aneuploid (some unusual combination not necessarily divisible by 23). But let's get back to all the other cells that we have, which are diploid. A normal diploid cell has two doses of each gene, one on each of the two chromosomes present. As an example, let's use the gene for the protein that when expressed gives rise to eye color. A person may have two copies of the gene for brown eyes, two copies of the gene for blue eyes, or one of each. Those two copies of the gene are the alleles of the gene. Genes exist in potentially different forms on the two chromosomes present and are said to exist as alleles (or forms) of that gene. Using our eye-color example, someone with two alleles of the same gene (two brown or two blue) is said to homozygous for the presence of that gene. Someone who has one of each allele (one brown and one blue) is said to be heterozygous for the presence of that gene. [☞ also Genotype; Phenotype]

Anneal

DNA, which is normally double-stranded in nature, can be manipulated in the laboratory in a number of ways to make it single-stranded. This process, known as denaturation, can be done by heating DNA to near boiling temperatures or treating it with a strong alkali. This process is done in order to ask a question about a particular DNA preparation, such as "Is an infectious organism's DNA present in the mixture?" or "Does this patient's DNA have a particular genetic mutation?" Molecular pathologists go about answering these kinds of questions by using a small piece of DNA (known as a probe) which is complementary to the target being sought, i.e., the microorganism or the mutation. [☞ Complementary Strands of DNA; Denature; Probe] In other words, the probe has the right matching sequence to seek out and find the target DNA sequence of interest. The biochemical process of binding a probe to a target by joining two pieces of complementary DNA is known as annealing.

In the example cited, a special form of annealing called hybridization has occurred because a hybrid DNA duplex has been formed. Under other laboratory conditions, the investigator might denature and then reassociate the DNA strands (no probe is involved) under conditions that favor that, by manipulation of things such as temperature or ingredients in the buffer in which the DNA is contained; this is the general process of annealing. In Polymerase Chain Reaction (PCR), and other *in vitro* nucleic acid amplification technologies, primers anneal to complementary DNA sequences as part of the overall performance of those reactions. [☞ also PCR; Primers]

Anticodon

Anticodons are three nucleotide long sequences that are specific for a target in messenger RNA (mRNA). Let's back up. DNA is transcribed into mRNA, which is translated into proteins. [☞ Gene Expression; Genetic Code] The sequence contained in mRNA, which was dictated by the DNA, or gene, that coded for it, is translated by the cellular machinery into proteins. Protein synthesis inside the cell is a rather complicated biochemical process. In short, in happens inside a protein synthesizing "machine" called a ribosome. The ribosome is where mRNA and amino acids come together in a specific way and the protein that is encoded by that mRNA molecule elongates until it's done. The amino acids in the cell (which got there because you had a burger for lunch, or a glass of milk, or a slice of cheese on your tuna sandwich) get to the ribosome for protein synthesis because they are commandeered by a molecule that takes them there; that molecule is called transfer RNA (tRNA) and looks something like a cloverleaf.

There are specific tRNA cloverleafs for specific amino acids. The amino acid binds on top of the cloverleaf and on the bottom is a three base-pair sequence called an anticodon. Based on the laws of complementarity [☞ Complementary Strands of DNA], the anticodon binds to the specific sequence in the mRNA associated with the ribosome that happens to code for, say, amino acid leucine. There is a specific codon in mRNA for leucine and a specific anticodon on the bottom of a leucine tRNA cloverleaf that recognizes the sequence, binds there, and gives up the leucine at the top of the tRNA cloverleaf to the growing protein chain in the ribosome.

Antiparallel

The two strands in a DNA double helix are antiparallel to each other. Chemically speaking, each strand or chain is made up of repeating units of deoxyribonucleotides, each linked to the next. Deoxyribonucleotides are composed of phosphate groups, a pentagon-shaped sugar molecule, and nitrogen-containing bases. Each of the positions in the sugar molecules is numbered, and the phosphate groups serve as chemical bridges attaching each nucleotide to the next. These phosphate bridges link the 3 position of the sugar in one nucleotide to the 5 position of the next, so that one strand runs 3-5-3-5-3-5-, etc., and the other strand runs in the opposite direction: 5-3-5-3-5-3-, etc. The two strands are said to be antiparallel to each other due to their chemical structures.

Autoradiograph

An autoradiograph, or autorad for short, is like an X-ray that you get to find a tooth cavity or a bone fracture. An autoradiograph is the end result of the Southern blot. [☞ Southern Blot] When a Southern blot is performed, one searches for a particular gene or gene fragment buried within all the DNA purified from a patient specimen. This is very much akin to looking for a needle in a haystack—or, more appropriately, looking for a single piece of hay within a haystack. The process generates a band-shaped or dot-shaped image that denotes that we found the needle—the gene fragment—of interest. Because the DNA probe that we used to find the genetic target has been radioactively labeled, when we place a piece of X-ray film atop the piece of heavy-duty nylon-type paper that is the Southern blot, that radioactive hybrid (target DNA plus radioactive probe) exposes the film. When we develop the X-ray film by standard film-developing methods, what is generated is called an autoradiograph: "auto" because it exposed itself, "radio" because it uses radioactivity, and "graph" because it's something like a photograph. Visual inspection of the autoradiograph allows us to answer questions about the presence of some aspect of that patient's DNA that might be instructive in making a particular diagnosis. When one uses a probe that is not radioactive, but rather luminesces under the right conditions, the resultant film is called a "lumigraph." [☞ Chemiluminescence]

Avery, Oswald T.

In 1944, Dr. Avery and his colleagues, C. M. MacLeod, M. McCarty, and their co-workers, performed a series of classic experiments with strains of bacteria that cause pneumonia *(Diplococcus pneumoniae)* and showed that DNA is a carrier of genetic information. These experiments are considered historical landmarks in the field of biology.

B-DNA

The naturally occurring form of DNA inside cells. B-DNA has the normal shape, angles of binding for the nucleotides that form DNA, and distances within the double helix that are found within DNA in the body or in solution in the laboratory. [☞ also A-DNA]

bDNA

This is not the second team, but rather is branched DNA. bDNA is the basis of a PCR "wannabe" [☞ PCR "Wannabes"]. It is an *in vitro* nucleic acid amplification technique in which the signal is actually amplified as opposed to the target. A fair amount of biochemistry occurs; if the target in the sample is present (the target is usually a beastie like the Hepatitis C virus or Human Immunodeficiency Virus), then a reaction that generates detectable quantities of light occurs and the test is positive. If the virus is not in the initial patient specimen, no light is generated and the test is read as negative. The test was developed and is marketed by Chiron Corporation in Emeryville, California.

Bacteriophage

Phage is another word for virus. Viruses don't just prey on humans; even lowly bacteria have to be careful about whom they conjugate with. A bacteriophage is a virus that specifically infects different kinds of bacteria. Examples include lambda (λ), T4, and Qβ. Scientists have learned much about gene expression, in general, by studying the simple genomes of these organisms and their life cycles within their bacterial hosts.

Band

There were some great ones when I was growing up in the '60s and '70s, but I suspect you'd rather read about the kind that relates to DNA. When DNA is electrophoresed in order to study it, it is placed into an electrophoretic gel well which is shaped like a rectangle. As electrophoresis proceeds to completion, DNA fragments are separated and can be visualized [☞ Ethidium Bromide]. The DNA fragments retain the basically rectangular shape of the well, but have been condensed by the process of electrophoresis into a tight line of visible DNA, referred to as a band. Band-shaped images are also generated by the detection phase of Southern blotting. [☞ Autoradiograph; Electrophoresis; Southern Blot]

Base Pairs

DNA is made up of strands of nucleotides which pair up with each other according to rules of complementarity. [☞ Complementary Strands of DNA] There are 3,000,000,000 (3 x 10^9) nucleotide base pairs in a human sperm or egg cell. Non-sex cells (somatic cells) like stomach, nerve and muscle cells have twice as much DNA. Here are some fairly useless arithmetic facts:

If you multiply the number of base pairs in a somatic cell (6 x 10^9) by the length of the DNA purified from a single cell and string out that DNA in a straight line (3.4 $\times$ 10^{-10} meters per base pair) the product is about 2.04 meters of DNA per cell. If you multiply 2.04 meters of DNA in one cell by the number of cells in a mature adult human (about 3.5×10^{13}) the product is 7.14×10^{13} meters. (714 is also the number of home runs Babe Ruth hit in his major league career.) The distance from the Earth to the Sun (one way) is about 93,000,000 miles. One mile is about 1625 meters, so the distance to the Sun is about 1.5×10^{11} meters. If you divide 7.14×10^{13} by 1.5×10^{11}, you find that the number of meters of DNA in one person could be strung back and forth between the Sun and the Earth 476 times. That's a lot of DNA.

Bases

We're not talking baseball here, we're talking freshman biochemistry. DNA and RNA are made up of bases which are ring-shaped chemical structures composed of carbon, hydrogen, nitrogen, and oxygen in various combinations. DNA is made up of the bases adenine, guanine, thymine, and cytosine (A, G, T, and C, respectively). The bases in RNA have an extra oxygen molecule and include A, G, C and uracil (U). There are no thymine bases in RNA. [☞ Nucleotide]

BRCA1

The *BRCA1* gene was cloned late in 1994. [☞ Clone] It is this gene which, when mutated, is responsible for a fraction of heritable breast cancer (as opposed to sporadic breast cancer, the overwhelmingly dominant form of the disease). There are particular populations in which *BRCA1* mutations are prevalent and mutation screening may be appropriate. But how do we proceed?

Individuals with a family history of breast cancer may pressure physicians for this test. If the patient is shown by *BRCA1* testing to harbor the same mutation as an affected primary relative, then there is reason for concern: The patient has an increased lifetime risk for breast cancer, although the disease can strike at any time, at age 30 or age 80. The medical community simply does not know whether it is best to offer such a patient a prophylactic double mastectomy. Furthermore, there is

9

the associated increased risk for ovarian cancer, raising the question of further prophylactic surgery. Even surgery could leave behind some cells that harbor the mutation. The patient and physician may decide that more frequent mammograms are advisable, but we don't know enough about this gene and the mutations it harbors. The mutations may act by increasing susceptibility (and cancerous transformation) to the very kind of ionizing radiation generated during mammography. Admittedly, what we know about the gene now suggests that this is not the mechanism of action. It is possible that discovering a *BRCA1* mutation may result in diligence about mammography for the rest of a woman's life in the same way that we now take for granted the importance of closely monitoring cholesterol and lipoprotein (HDLs and LDLs) levels in people found to be at risk for heart disease. Such an eventuality, should it come to pass, would be a good thing for the overall health of the nation; everyone, including insurance companies, should come to realize these possibilities in time.

On the other hand, an individual with a relevant family history may be tested and shown not to harbor that specific mutation present in the family. This could cause the individual to proceed through life with a false sense of security, bypassing regular mammograms and ignoring dietary concerns. Inherited breast cancer represents a small fraction of all cases of breast cancer. Moreover, *BRCA1* is only one of the genes that may be related to increased risk of breast cancer. Within *BRCA1*, we now know of over a hundred mutations in this gene associated with breast cancer, and that number will certainly grow. Testing for one or a few and not finding them is no guarantee against breast cancer.

The cloning of *BRCA1* was a great scientific achievement. It has generated potentially exciting and useful clinical options. At the same time, it leaves us with many questions on the best way to proceed with this knowledge.

For Internet browsers, the Breast Cancer Information Core data base is accessible on the World Wide Web at:

http://www.nchgr.nih.gov/dir/lab_transfer/bic/

You may use this address to learn more about *BRCA1* and breast cancer:

http://www.nchgr.nih.gov/intramural_research/lab_transfer/bic/

10

Or just type in "breast cancer" as the keywords in an Internet search, use your Web Browser, and you will uncover many interesting and useful places to visit.

Other network databases of familial breast cancer are available through Online Mendelian Inheritance in Man and the Breast Cancer Information Clearinghouse.

For more information on breast health and breast cancer, contact:

National Cancer Institute's Cancer Information Service
☎ 1-800-4-CANCER.

The American Cancer Society
☎ 1-800-ACS-2345.

The Y-ME Hotline
☎ 1-800-221-2141.

National Alliance of Breast Cancer Organizations
☎ 1-800-719-9154.

cDNA

cDNA is complementary DNA, which does not mean that it is especially polite but that it is an unusual biochemical entity. (Sounds like Star Trek technobabble, doesn't it?) DNA is the target of a powerful laboratory method called Polymerase Chain Reaction (PCR). [☞ PCR] Sometimes we need to ask a question about RNA and not DNA. Examples include when we are searching in the diagnostics laboratory for the presence of a virus that naturally only contains RNA (for example, nasty buggers like the Hepatitis C virus or Human Immunodeficiency Virus), or if we are investigating not the presence of a gene (DNA) associated with disease but rather the expression of that gene as RNA. [☞ Gene Expression] In such cases, we still want to exploit the power of PCR, but we first need to purify the RNA of interest and turn it into DNA so that we can proceed with the PCR method in the laboratory. We do that with an enzyme called reverse transcriptase. [☞ also Retroviruses; Reverse Transcriptase]

We mix the purified RNA with Reverse Transcriptase (RT) and other necessary ingredients, and when this enzyme encounters RNA it turns that RNA molecule into a DNA "copy" of that RNA. That DNA is complementary to the RNA that the enzyme used as a template, and is therefore called complementary DNA, or cDNA for short. [☞ Complementary Strands of DNA] cDNA can then participate in PCR just like any other DNA molecule. In brief then: RNA + RT = cDNA. (An enzyme

12

called *Tth* polymerase has the ability to combine the activities of reverse transcription and the important enzyme in PCR, DNA polymerase, whose job is to make more DNA. *Tth* polymerase is an enzyme from the bacteria *Thermus thermophilus;* hence the name.)

Chemiluminescence

In brief, chemiluminescence refers to the emission of light from a chemical reaction. The phenomenon was first described in 1877. We now have the ability to label (through chemical attachment) DNA probes with chemical compounds that act as reporter molecules. When we hybridize such DNA probes to DNA targets of interest and then perform the necessary chemistry, light is emitted which reports to us a successful hybridization between target DNA (patient DNA in the clinical setting) and the probe we used. That emitted light can be captured on X-ray film, which we can study to help us answer questions about the nature of that patient's DNA. [☞ also Autoradiograph, keeping in mind that an "autorad" generated by a chemiluminescent probe is usually called a lumigraph.] Instruments called luminometers also exist that capture emitted light and give us information about the kinds of analyses described here.

Chemiluminescence occurs in nature too, where it is called bioluminescence. Examples are found in certain marine bacteria and the firefly.

Chromosomal Translocation

Chromosomal translocation is an abnormal occurrence. It is the exchange of portions of chromosomes one with another, which is specifically referred to as reciprocal translocation. Another type of chromosomal translocation, called centric fusion, involves two complete chromosomes fusing to each other. Some reciprocal translocations are known to be involved in the generation of certain cancers. The mechanism for this carcinogenesis involves the movement of genes during the translocation from one "address" to another where the gene has escaped its normal regulation by the cell, causing the uncontrolled growth that is cancer. Chronic myelogenous leukemia is such an example; it results from a balanced chromosome translocation, called the Philadelphia chromosome after the place where it was discovered. The Philadelphia chromosome can be detected by cytogenetic laboratory analysis, or more accurately and more often by molecular techniques such as the Southern blot and reverse transcriptase PCR. [☞ also Gene Rearrangement; PCR; Southern Blot]

Chromosome

Literally, "colored body", referring to 19th-Century scientists' microscopic observation of these blue and red staining materials. In primates, man, and in all higher organisms, DNA is contained within the cell nucleus in tightly packed structures called chromosomes. Chromosomes consist of DNA and proteins (virtually equal parts of histone and non-histone proteins). [☞ Histone] The proteins help package the DNA by serving as a kind of scaffolding so that the very long DNA molecules present can be condensed into a very small space.

Humans have 23 pairs of chromosomes in every cell (except mature red blood cells) that are visually distinct only during cell division, a process known as mitosis. When the cell is in that portion of the cell cycle where it is not dividing–the interphase period–chromosomes cannot be individually differentiated. Gametes, or sex cells (sperm and eggs) have half the normal complement of chromosomes, so that when they combine to form a fertilized egg, the full complement of chromosomes (and DNA) is present, which can then go on to form an embryo.

Number of Chromosomes in 1 Cell of Different Species

Bacteria	1	Cat	38
Fruit flies	8	Mouse	40
Peas	14	Rat	42
Bees	16	Rabbit	44
Corn	20	Human	46
Frog	26	Chicken	78
Fox	34	Some species	
		of fern plants	> 1000

Cistron

Another word for gene, seldom used.

CLIA '88

CLIA stands for Clinical Laboratory Improvement Amendments of 1988. CLIA '88 is a Federal Law describing, among other things, the necessary qualifications for clinical laboratory workers and directors; what must be done before a new clinical test is implemented in the laboratory; aspects of quality control, quality assurance, and proficiency of the laboratory; and much more. Clinical laboratory tests are subdivided within CLIA '88 into low, moderate, and high complexity; most DNA-based testing qualifies as high complexity.

Clone

(1) To clone a particular piece of DNA, including a gene, means to molecularly isolate it in the laboratory and insert it into a cloning vector. A cloning vector is another piece of DNA that can be inserted into a bacteria, virus, or yeast cell, which then grows, simultaneously making many, many copies of the inserted vector that has the piece of DNA that was isolated and placed into the vector. So to "clone" a piece of DNA means to isolate it and make more of it for study or use.

An example of a use for cloning is to molecularly clone the gene for insulin, insert it into an appropriate vector, introduce this so-called recombinant DNA molecule (cloned DNA plus vector) into cells that can be grown either in the laboratory or on some large scale. During growth the cells with the introduced insulin gene express that gene, thereby releasing insulin into the mix, which is purified and used medically.

(One of my favorite "scientific" cartoons depicts a little boy who brought a frog to "Show and Tell." When the teacher reminds him that he was supposed to have brought something that he had made, the boy calmly replies: "I cloned him.")

(2) A clone of cells is a group of cells which are genetically identical to each other. In leukemia, for example, one white blood cell may escape the normal growth regulation the body imposes on its cells, due perhaps to a mutation caused by high-voltage electric fields, overexposure to sunlight, or any one of a number of environmental insults. That cell becomes leukemic or cancerous and divides uncontrollably, causing disease. All of the daughter cells that arise from the original leukemic cell are genetically identical and form a so-called monoclonal population of cells.

Codon

A three base-pair sequence in DNA that codes for an amino acid. When DNA is subjected to transcription (into RNA) and then translation (into protein) by the cellular biochemical machinery that carries on these things, each single amino acid in the growing protein chain is coded for by a sequence of three bases in the gene (DNA) that coded for that protein. Those three bases are termed a "codon." [☞ also Anticodon; Genetic Code]

Complementary Strands of DNA

DNA is an awfully polite molecule; the two sister strands are always complementing one another. The double-stranded DNA helix is made up of bases (among other things) which follow strict rules of complementarity (note the spelling) that dictate how the strands pair up with each other. The rules are quite simple: the base adenine (A) always pairs in DNA with thymine (T); and guanine (G) always pairs with cytosine (C). Thus, the base pairing is A to T and G to C. (My biochemistry study partner in undergraduate school had a mental block about A-T and G-C and always wrote it on his hand to remember it. I should come up with a mnemonic so you too could remember, but I think it's really easier to just memorize it. But now you know how to spell mnemonic!) So base pair complementarity dictates the sequence of the sister strand (and of the daughter strand that is synthesized during DNA replication). If one strand is:

AGCTTTAAGTCGCTTA, then the complementary strand must be: TCGAAATTCAGCGAAT.

You may also deduce from the above that the following statement of DNA is true: Within DNA the number of guanine bases = the number of cytosines, and the number of adenines = the number of thymines; in other words, A = T and G = C.

In RNA, thymine is replaced by uracil (U). In RNA, which is a more or less single-stranded molecule, local regions of double-strandedness [☞ Denature] can occur and the base pairing is A to U and G to C.

Crick, Francis H. C.

Francis H. C. Crick and James D. Watson (with a little help from their professional colleagues) deduced the double-helical nature of DNA, realized how that structure lent itself to replication of the molecule, shared the 1962 Nobel Prize for their work (along with Maurice Wilkins), went on to publish many more scientific manuscripts, write books, give talks, become faculty, and head scientific research institutes. For their work, scientists affectionately call one strand of double-stranded DNA the "Watson strand" and the other strand the "Crick strand."

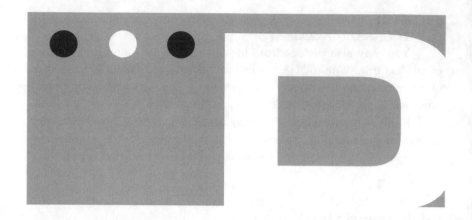

Denature

DNA is naturally double-stranded. Often in the laboratory, in order to work with DNA or detect a specific feature of it (for example, a mutation), we must make the DNA single-stranded so that we can get at the specific sequence of interest. To "denature" DNA is to make it single-stranded. This is most often done by heating the DNA solution to boiling or to temperatures near 100° C, or by treating the DNA with strong alkali.

RNA is a single-stranded molecule, but it may exhibit local regions of double-strandedness. This can occur due to the particular base sequence in a given RNA molecule; the sequence may be such that, under the right chemical conditions, some base pairing occurs, creating double-strandedness in an otherwise single-stranded molecule. Under these circumstances, it may be appropriate to denature RNA to work with it further in the laboratory.

DNA (Deoxyribonucleic Acid)

"The stuff of life." DNA is the genetic material that is passed from parent to progeny and propagates the characteristics of the species in the form of the genes it contains and the proteins for which it codes. The photograph below is that of purified human DNA, hanging from the end of a glass rod (the liquid you see is alcohol). [☞ Ribonucleic Acid (RNA); also Nucleotide; Nucleic Acid; in fact, every entry in this book.]

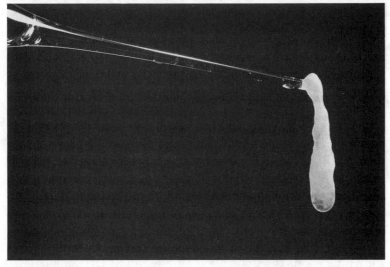

Photo by Peter Roberts

DNA Bank

This is a service just like a sperm bank, tissue bank, or financial institution where money is stored. Many institutions, including the hospital where I work, have DNA banks where DNA extracted from certain patients' tissues (at the patient's request with medical advice, of course) is frozen and stored indefinitely. In this way, the DNA is available if for some reason it needs to be tested in the future. The reasons for such testing are not the most pleasant circumstances to consider. For example, matching DNA profiles obtained from remains with banked DNA on a known individual can provide a basis for absolute identification. In fact, this has been exploited by the military; the Gulf War was the first in our history to have no interments in the Tomb of the Unknowns in Arlington National Cemetery.

From a medical point of view, genetic disorders may be difficult or impossible to diagnose. Children may present to clinics with rare, unrecognized, or unique ailments. A clinical diagnosis may be suspected, but insufficient laboratory data may be available upon which to base a definitive diagnosis. No confirmatory lab test may be available for the condition; such patients may be provided with a definitive diagnosis by future research and future DNA diagnostics tests. The banking of DNA from these affected patients, who may not survive, may permit definitive diagnosis and recurrence risk counseling for the parents of these patients. The availability of such DNA may also be of value in the counseling of siblings and other family members.

The value of DNA banking is not limited to such unfortunate circumstances. A stored DNA sample of a parent or grandparent may be of value to descendants of that individual with respect to counseling concerning reproductive and health issues. [☞ Paternity/Profiling/Identity/Forensic Testing by DNA]

DNA Chips

I don't believe we'll be seeing this as the latest offering from Frito-Lay anytime soon. DNA chips are an attempt at DNA miniaturization, which is ironic (isn't it?) because DNA is already so small. DNA chips are ordered arrays of oligonucleotides ("oligos" for short). [☞ Oligonucleotide] The oligonucleotides are bound chemophotolithotropically (now that'll really impress 'em at the Monday morning staff meeting: it means combining light-directed chemical synthesis with semiconductor based photolithography and solid phase chemical synthesis; trust me, the point is, they get it done) to solid phase support chips. Oligos of defined length can be attached to the solid support in a variety of patterns. For example, 256 different oligos eight base-pairs long (8mers) can be attached on a 16 by 16 format. Or you could really scale it up and attach to the support enough oligos of enough length to encompass all the possible sequences in a given gene of interest. Then react that with a particular patient's DNA and detect the areas of perfect matching (by some light-generating detection system) to learn whether the patient has a particular mutation. This is exciting work that is being carried out at Affymetrix in Santa Clara, California, and may become a highly automatable way to learn about individual DNA sequences in the not-too-distant future.

DNA Extraction/Purification

The process of purifying DNA from tissue. That tissue could be whole blood, solid tissue, bone marrow, cerebrospinal fluid, etc. Every cell in the body is nucleated (except mature red blood cells) and contains DNA; therefore, every tissue is suitable for DNA extraction. Blood is obviously an excellent tissue source for DNA due to its accessibility and the fact that it is full of white blood cells and lots of other DNA-containing cells that aren't red blood cells.

RNA extraction, using different methods and chemicals, is also routinely done in the clinical molecular pathology laboratory.

DNA Fingerprinting

[☞ Paternity/Profiling/Identity/Forensic Testing by DNA]

DNA Labeling

DNA molecules can be "labeled" through a variety of biochemical manipulations that rely on the action of proteins or chemical modification. "Labeling" involves tagging a DNA molecule with some reporter molecule that can be detected by using an X-ray film or more chemical reactions that generate visible color. What we see on the film or we view by means of observing color allows laboratorians to answer questions about a particular sample of DNA. Examples of these questions include: "Is a particular microorganism's DNA present in this patient sample?" "Is a particular mutation present in this DNA sample?" "Is there a particular genetic marker present that indicates the presence of cancer, or indicates paternity?" [☞also Autoradiogram; Chemiluminescence; Southern Blot]

DNA Probe

[☞ Probe]

21

DNA Sequencing

In the mid to late '70s and on into the early '80s, we invented and refined techniques to actually read the sequence present in a particular piece of DNA. In the mid-1990s we have scaled that up to the point where we are making excellent progress on sequencing the entire three billion bases in the human genome. DNA sequencing techniques are based on DNA electrophoresis that is done in high-resolution polyacrylamide gels (not unlike agarose gels in principle), also called sequencing gels. Sequencing gels are capable of resolving single-stranded oligonucleotides hundreds of base pairs in length that differ in size by just a single deoxyribonucleotide. Through enzymatic or chemical reactions, oligonucleotides encompassing the region of interest are made that end with either adenine (A), guanine (G), thymine (T), or cytosine (C)–the four deoxyribonucleotides that make up DNA. The oligonucleotide products of the reactions are then electrophoresed in adjacent lanes of a sequencing gel. The one-base-pair resolution capability of these gels allows us to "read" from the gel the sequence of the DNA under analysis.

There are two general methods for performing DNA sequencing: dideoxy (Sanger) sequencing and chemical (Maxam-Gilbert) sequencing. In dideoxy sequencing, 2', 3' dideoxyribonucleotide triphosphates (ddNTPs) are used as substrates for growing oligonucleotide chains synthesized from the DNA of interest as a template. When a ddNTP is incorporated, oligonucleotide chain growth is blocked because that chain now lacks a 3' hydroxyl group for continued chain elongation. Four separate reactions are run, each with a unique ddNTP. Manipulation of the ddNTPs:dNTPs ratio results in chain termination at each base occurrence in the DNA template corresponding to the included ddNTP. In this way, populations of extended chains exist within each reaction that have differing 3' ends specifying a given ddNTP–in other words, specifying the sequence. In chemical sequencing, radioactively labeled DNA undergoes reaction with chemicals that specifically cleave at certain bases. DNA sequence is determined directly following electrophoresis and autoradiography.

From the molecular point of view, DNA sequencing is the "gold standard" for detection of mutations and relevant DNA sequences. From a practical and clinical point of view, however, automated DNA sequencing equipment is extremely expensive, and DNA sequencing will not become routine until equipment becomes less costly and more user-friendly, and until it has been demonstrated that the benefits provided by sequencing can have a positive impact on patient management and disease outcome. In other words, we need to know what a particular DNA sequence change means to the patient and his or her treatment; finding this out will require a lot more clinical research. DNA sequencing has been relevant, both directly and indirectly, in virtually every disease diagnosable at the molecular level, including cystic fibrosis, tumor suppressor gene analysis in cancer, and much more. I say "indirectly" because even though a particular disorder may be detectable by Polymerase Chain Reaction (PCR), the success of a particular PCR depends on knowing the sequence of the gene of interest so that appropriate primers for PCR can be synthesized [☞ also Agarose; Electrophoresis; Human Genome Project; PCR; Primers]

DNase

[☞ Nuclease]

Dot Blot

[☞ Southern Blot]

Duplex

Depending on where you learn your real estate, duplex refers to two homes or domiciles that are one on top of the other or side by side. In DNA chemistry though, duplex DNA refers to the normal state of affairs, that is, good old double-stranded DNA.

Electric Genes

Double-stranded DNA can conduct electricity, in the form of electrons, under the right laboratory conditions. Single-stranded DNA can do the same but conducts at a much slower rate. Tom Meade and Jon Kayyem at the California Institute of Technology in Pasadena have exploited this finding and are continuing their work to develop rapid, electron-based DNA diagnostic tests. The possible medical applications are exciting and depend upon the measurable differences in velocity described above. However, I honestly can't recommend that you call your stockbroker just yet.

Electrophoresis

Electrophoresis is a very commonly used laboratory technique, both in the clinical laboratory and the research laboratory. It is a technique that takes advantage of the fact that molecules like DNA and protein migrate in an electric field. DNA migrates in an electric field inversely proportional to its molecular weight. That's a fancy way of saying that the heavier (or larger) a piece of DNA is, the more slowly it migrates, while the lighter (or smaller or less massive) a piece of DNA is, the more quickly it migrates.

Typical agarose DNA electrophoresis in the laboratory proceeds like this: You pour a molten gel [☞ also Agarose], let it cool into a semi-solid material, overlay it with water to which has been added the right salts and chemicals (called electrophoresis buffer), and then load your DNA solution (you may or may not have cut the DNA into fragments). [☞ also Restriction Endonucleases] Mixed into your DNA solution are a dye (generally blue) and sugar (sucrose). The blue dye accomplishes two things: it allows you to see what you're doing and it allows you to monitor the progress of electrophoresis. The sucrose added to the DNA solution makes it more dense than the water-based buffer with which you've overlaid your gel. Because of this added density, your DNA solution stays in the well in the gel that you're loading with DNA. Without the sucrose, you'd be adding a small amount of water-based DNA solution to a gel overlaid with lots and lots of water-based buffer, and your DNA would go off and wander away into solution and you'd feel like cursing and swearing.

Now that you've successfully loaded the gel, you attach electrical leads to each side of the gel box and to a power supply, turn on the electricity, usually in the range of 20 to 250 volts, and now you can usually go home or do something else in the laboratory. Electrophoresis can take from as little as 10 or 15 minutes to overnight, depending on what you're trying to accomplish. The reason this works is that you have created an electrical circuit between the gel and the power supply, and as electricity flows through the buffer you used to overlay the gel, it carries along the DNA molecules you loaded. You can watch the progress of electrophoresis by watching the blue dye, but that's only a little bit more exciting than watching paint dry.

The blue dye also serves another purpose, which is to make sure you didn't reverse the polarity when you plugged in the electrical leads. If you make that error, both the DNA and the dye will go in the wrong direction and your work will be ruined. We usually come back a couple of minutes after starting electrophoresis to make sure we haven't committed the dreaded laboratory error of "retrophoresis." If we have, we turn off the power and change the leads to the correct position, confident that electrophoresis will proceed properly. Then we can go home.

DNA electrophoresis has a wide variety of uses, including DNA sequencing, DNA fingerprinting, DNA quality assessment, and DNA restriction fragment analysis.

Enhancer

Enhancers are stretches of bases within DNA, about 50 to 150 base pairs in length, that increase the rate of gene expression. Enhancers have stretches of bases in them that are recognized and bound by different DNA binding proteins. These proteins act in different ways to regulate the expression of genes. Enhancers may be physically close to, or far from, the gene they are responsible for regulating.

Enzyme

Enzymes are the tools of molecular biology. Enzymes are proteins (encoded by genes) that catalyze a biochemical reaction. In other words, enzymes make biochemical reactions occur. Enzymes carry on the business of life, whether that is digesting the building blocks of the food we eat, making more DNA, carrying oxygen molecules along the necessary path so that our cells can use that oxygen, or countless other examples. We have learned how to purify enzymes from natural sources and to use them as tools in the laboratory. Some of the most commonly used enzymatic tools in the molecular biology laboratory are restriction endonucleases, DNA polymerase, and reverse transcriptase (notice that enzyme names always end with the suffix "-ase"). When we mix DNA or RNA, under controlled conditions, with different enzymes and necessary ingredients chosen to accomplish a specific task, we manipulate DNA or RNA in order to learn more about it and find any clues to the disease we might be investigating in the clinical molecular pathology laboratory.

Ethidium Bromide

A commonly used dye. Ethidium bromide (EtBr) is a chemical whose structure contains, in part, a hexagonal carbon ring. Imagine this ring structure in a flattened plane; as such, it can insert itself (the technical term is "intercalate") between the bases that make up the DNA double helix. Once inserted, EtBr changes the physical characteristics of DNA such that when EtBr-stained DNA is illuminated with ultraviolet light (those "black lights" we had in our rooms as teenagers in the '60s were more scientific than we knew), it fluoresces. Fluorescent, EtBr-stained DNA is easily detected and amenable to photography so that a permanent record can be made. With this explanation in mind, it is not surprising to learn that EtBr is a mutagen and is something that must be worked with carefully in the laboratory.

Exon

No, I didn't misspell the name of the oil company (but if you want to use that little device to remember this, that's OK with me; I've got a better device for this later on—keep reading). Genes are made up of DNA which is transcribed into messenger RNA (mRNA) and then trans-lated into proteins which carry on the "stuff of life." [☞ also Gene Ex-pression] The DNA in a gene is arranged in a section-like fashion. There are alternating stretches of DNA that do and do not code for the ulti-mate gene product, the protein. The sections of the gene that are ulti-mately translated into a part of the protein are called exons; exons are expressed. The intervening stretches of DNA in between exons are called introns; they are spliced out of the gene when it is made into RNA and serve as regulatory parts of the DNA, or punctuation marks. Some scientists have referred to introns as "junk DNA." [☞ also Intron; Junk DNA; Gene Splicing]

Expression

[☞ Gene Expression]

Extension

What you ask for on April 14 when your taxes are nowhere near done. Or, with respect to DNA, extension refers to elongation of the growing DNA chain that is being synthesized using the parent DNA strand as the template for synthesis of that daughter strand. This is a natural process that occurs during DNA replication. It is also a process that scientists have learned to mimic in the laboratory using different re-agents and polymerase enzymes (proteins whose job is to synthesize new strands of DNA or RNA) to artificially create new DNA, something akin to making a haystack full of needles. [☞ PCR]

Forensic DNA Testing

[☞ Paternity/Profiling/Identity/Forensic Testing by DNA]

Fragile X Syndrome

Fragile X syndrome is the most common form of hereditary mental retardation in males. The frequency of the disorder is one in 1,000 to 1,500 individuals. In addition to mental retardation, Fragile X Syndrome is also associated with characteristic clinical symptoms including developmental delay, long and prominent ears, high arched palate, prominent jaw, long face, hyperextensible joints, hand calluses, characteristic behavioral and neurological difficulties, double-jointed thumbs, single palmar crease, flat feet, macroorchidism (overly large testicles), and more.

The disorder got its name from the way in which it used to be diagnosed in the clinical laboratory, usually in the cytogenetics laboratory. A blood specimen was taken from the patient and the purified blood cells were grown in a way such that laboratorians could actually observe that a site on the X chromosome of these cells could be induced to break due to the presence of a particularly fragile site. We now know that, in

29

affected patients, the site is fragile because of an unusual characteristic of the DNA there.

Unaffected individuals have at this site of DNA a series of three nucleotides–cytosine-guanine-guanine, or CGG–which is repeated anywhere from 6 to 52 times. In affected individuals, that CGG trinucleotide is repeated over 200 times, sometimes extending into the thousands. This destabilizing feature of the DNA helped explain why a break could be introduced there when these cells were grown in the laboratory. Individuals with between 52 and 200 repeats of CGG are in the carrier category for the disorder and have no symptoms of Fragile X Syndrome. In the clinical laboratory we also look at special side groups on the DNA to give us insight about carrier versus affected status.

This form of mutation, called trinucleotide repeat amplification, has been shown to occur in several other diseases including Huntington Disease and Myotonic Dystrophy. In Fragile X Syndrome, the introduction of all those CGGs interferes with normal expression of the gene there, and it is that lack of expression of an important gene that leads to the clinical symptoms or Fragile X phenotype. [☞ Phenotype]

Since this DNA abnormality was found in the early 1990s, a direct DNA-based test to detect Fragile X Syndrome has been in use and represents a faster, cheaper, more specific, and more sensitive way to diagnose the disorder than was previously available through cytogenetic analysis. At the same time it is important to remember that the DNA-based Fragile X test is highly specific for that disorder, and routine cytogenetic analysis can turn up other abnormalities that would not be detected by molecular Fragile X analysis.

To learn more about Fragile X Syndrome, talk to your physician or genetic counselor, or contact:

The National Fragile X Foundation
1441 York Street, Suite 215
Denver, Colorado 80206
☎ 1-800-688-8765 or 303-333-6155

GC-rich

Genes–indeed, all DNA–are composed of nucleotides: guanine (G), cytosine (C), adenine (A), and thymine (T). [☞ Nucleotide] When a particular stretch of DNA is particularly high in GC content, it is said to be GC-rich. GC-rich regions of DNA can be particularly troublesome to deal with when using them experimentally or as the target of a DNA diagnostic test.

GenBank

GenBank is a huge National Institutes of Health (NIH) genetic database comprised of known DNA sequences collected from scientists worldwide. It is administered and maintained by the National Center for Biotechnology Information (NCBI). As of June 1996, GenBank has approximately 552,000,000 bases in 835,000 sequence records. GenBank is part of the International Nucleotide Sequence Database Collaboration, comprised of the DNA DataBank of Japan (DDBJ), the European Molecular Biology Laboratory (EMBL), and GenBank at NCBI. These organizations exchange data daily. You can access GenBank on the World Wide Web at:

http://www.ncbi.nlm.nih.gov/Web/Genbank/

Gene

A gene is a segment of DNA with a specific architecture (start signals, stop signals, embedded regulatory elements, and more) that the cellular machinery recognizes and transcribes into RNA. Most genes are eventually translated into proteins that carry out the business of life. Some genes carry out their end function as RNA molecules (for example transfer RNA, or tRNA, molecules). Genes are responsive to different stimuli. For example, some genes may be turned off until a particular hormone interacts with them to turn them on so that they can express themselves. It's also a good thing that the genes that code for the proteins that make a stomach cell a stomach cell are not in the "on" position in a liver or brain cell. [☞ Tissue-Specific Gene Expression] Environmental insults such as cigarette smoke or high-voltage electric fields can cause mutations in some genes. The genes in elementary organisms like bacteria have a different, more simple architecture than genes present in organisms like plants or man. [☞ also Gene Expression]

Gene Expression

A close friend's brother-in-law (B. G.) and I once had a discussion. B. G. is a professional actor who is sensitive to the idea of "expressing oneself." He found it interesting and comical that we who work with DNA speak of gene "expression" and wondered just what does that mean, anyway. Well, here goes:

You may have heard the cliché, "DNA is the stuff of life." What this cliché means is that genetic information flows from parent to child through the DNA. The business of carrying on life (cellular biochemistry) in each cell, organ, and organ system in our bodies is carried out by proteins. Proteins are responsible for the color of our eyes, for ensuring that the oxygen we breathe is transported to the appropriate place in the cell for utilization, for the elasticity of our skin, for transporting and digesting nutrients, for our immune response to the cold virus our kid brought home from kindergarten . . . all that our bodies do to carry on life is mediated through the action of proteins. Individual proteins are invisible (to the naked eye) structures in our bodies that are generated by machinery inside our cells. The machinery is composed of still more proteins (we could probably get into a chicken-and-egg thing here, but that's beside the point) that carry on the business of protein synthesis. Proteins are synthesized in cytoplasm, the gelatinous goop (pretty scientific, eh?) inside our cells which surrounds the cell's nucleus where the DNA lives. Proteins are made of long stretches of fundamental molecules called amino acids. A protein might be composed of a few or many thousands of amino acids. How does the protein synthesizing machinery of the cell's cytoplasm know how to assemble in the correct order a given set of amino acids to form protein X, Y, or Z? The answer lies in the RNA.

RNA is the intermediary between proteins and DNA. Based on the sequence (of bases) in the DNA, an RNA transcript (loosely speaking, a transcript is a "copy") is generated. That RNA copy contains the instructions given to it by the DNA when the RNA was made off of the DNA template. Those instructions are faithfully read by the protein synthesizing machinery of the cell. (One way to cause mutations is to read those instructions unfaithfully.) Reading those instructions means translating the code in the RNA from bases (the building blocks of DNA and RNA) to amino acids (the building blocks of proteins).

So, in summary: the flow of genetic information is as follows:

DNA → RNA → Protein. The bases in the DNA are transcribed into an RNA intermediary (in the cell's nucleus) whose bases are translated into amino acids and ultimately proteins (in the cell's cytoplasm).

And as so often is the case when you're trying to learn something, there is an exception. In this case, the exception is reverse transcriptase. [☞ Reverse Transcriptase; Retroviruses; ☞ also Anticodon; Ribosomal RNA; Tissue-Specific Gene Expression.]

Gene Product

A gene product is the end result of transcription or translation. DNA is transcribed into messenger RNA (mRNA), which is then translated into protein. Sometimes DNA codes for an end product that is

RNA and not protein–for example, when DNA codes for the RNA that is one of the building blocks of ribosomes or transfer RNA molecules. [☞ also Gene Expression; Transcription; Translation]

Gene Rearrangement

One of my colleagues thinks of this as "gene tampering," which conveys artificiality, something that we in the laboratory barge in and do to the DNA. My colleague is not far off, except that the key difference is that gene rearrangement is natural gene tampering, something the cell does on its own to the DNA contained in it.

Gene rearrangement is a natural phenomenon in those species that are able to mount an immune response–including humans, of course. A large part of our immune response is dependent upon the production of antibodies. Even with all the DNA we have present in our cells, there is still not enough DNA to code for all the antibodies that are needed to deal with the many antigens that are present in the environment. (An antigen is anything that elicits an immune response, such as the influenza virus, ragweed pollen, or countless other examples.) That is because every *successful* immune response depends on the production of unique antibodies (more on that later) to *specifically* interact with and help defeat the invading antigen.

Antibodies are proteins, so their structure is encoded in our DNA. Higher species have evolved a way to deal with this information content or size problem described earlier. The genes that code for the proteins that make up our antibodies are arranged in a unique way. They are composed of many different segments or regions; you can think of them as cassettes of DNA coding information. In response to a particular immunological insult, our DNA shuffles around these cassettes in different ways. There are so many cassettes and so many different ways in which they can be shuffled, or rearranged, that unique antibodies can be made for the purposes of specific immunological interaction with an antigen. This so-called gene rearrangement is the way that we generate the necessary antibody diversity to deal with the vast number of antigens in the environment.

I mentioned above that every successful immune response depends on the production of unique antibodies; that's not strictly true. There are immune interactions that are cell-mediated as opposed to antibody-mediated. In the same way that our antibody coding genes rearrange, so too do the genes that code for the protein receptors on our immune cells; like T cells, for example (a subset of T cells is the target of infection by HIV). These so-called T cell receptor proteins are the ones that mediate the interaction between certain antigens and the T cells that help to fight them off.

So gene rearrangement is a normal process. We can take advantage of it to diagnose an abnormal process like leukemia or lymphoma. In these diseases, large numbers of a particular clone [☞ Clone, definition 2] of immune cells (B or T cells) have all rearranged an antibody gene or T cell receptor gene in the same way. Furthermore, this clone of cells is present at an unusually high, disease-causing number. It is the presence of that unique, normal gene rearrangement (think of it as a molecular signature unique for that clone of abnormal, cancerous cells) that can be detected in the molecular pathology laboratory to help in the diagnosis of certain kinds of leukemia or lymphoma. This test, called the B and T cell gene rearrangement test, is useful not only in initial diagnosis but also in monitoring the success or failure of therapy and in determining whether the return of disease, should it occur, is due to the same cancerous clone of cells or to a different one—information important to oncologists.

While the kind of gene rearrangement described above is normal, sometimes genes are rearranged due to an abnormal event. Chromosomal translocation is such an event. Chromosomal translocation is the abnormal exchange of pieces of chromosomes between each other—for example, a piece of chromosome 9 breaking off and attaching to chromosome 22. When this happens, the genes present on the piece of the chromosome that broke off are translocated—moved to a new address—inside the nucleus; they can also be thought of as having rearranged, this time as part of an abnormal process. This event occurs in several kinds of cancer and can be detected by a variety of tests available in the clinical molecular pathology laboratory. [☞ also Chromosomal Translocation]

Gene Splicing

Splicing is what my brother, Steve, used to love to do with his reel-to-reel tape recorder back in the '60s. He'd spend hours cutting and mending tape–splicing it to create new recordings. Actually that is the perfect analogy for how the term splicing is used with respect to DNA and RNA. The DNA in a mammalian, human, or plant gene is longer than the messenger RNA (mRNA) molecule that is derived from that gene. That's because of the presence of introns and exons. [☞ Exon; Intron] The introns are spliced out to form the mature mRNA sequence. The site between exon and intron in a gene is known as the splice junction. Splicing at the splice junction must be precise to the base because an error of even one base can disturb the reading frame of the resultant mRNA such that a mutant protein will be produced. In fact, splice site mutations are a class of mutations.

Gene Therapy

Many diseases are caused by a genetic malfunction. Large amounts of money and effort are being invested into research to correct malfunctioning genes. This is known as gene therapy. Successful gene therapy can take the form of introduction of a functional gene into a patient's cells so that expression of the corrected gene will reverse the defect caused by the abnormal gene. Such genetic correction needs to be tissue-specific in order to accomplish its task. For example, correcting a mutation that causes cystic fibrosis will likely need to occur in lung and pancreatic tissue–organs affected by cystic fibrosis–to reverse the improper function in those organs that causes symptoms.

Gene therapy can also be defined as introduction of a new function into a cell that is not strictly the introduction of a new gene. For example, cancer cells can be artificially immunostimulated by genetic mechanisms to help "vaccinate" a patient against his or her own tumor.

For ethical reasons, gene therapy is only done on somatic cells–those cells in the body that are not gametes (sperm or eggs). Genetic manipulation of gametes is unethical, for obvious reasons, and is not done by responsible scientists and researchers. Dozens of approved gene therapy clinical trials are currently ongoing in North America, Europe, and elsewhere for different genetic diseases, infectious diseases (especially HIV-1 infection), and cancer.

The clinical molecular pathology laboratory of the future will not only be a diagnostics laboratory. When gene therapy to treat human disease becomes a reality, the molecular pathology laboratory will be charged with identifying missing or damaged genes in patients to identify them as appropriate candidates for gene therapy. Furthermore, therapeutic agents composed of DNA and RNA will need to be monitored for degradation and purity. Genes newly introduced into patients will have to be assessed for proper insertion and demonstration and quantitation of new gene expression. Gene therapy represents the future and the promise of a new era in clinical medicine.

Genetic Code

A code is a series of items, words, symbols, or the like that make no apparent sense until that code is broken or solved so that it can be read. DNA is the same way. DNA is made up of nucleotides–guanine (G), cytosine (C), adenine (A), and thymine (T)–which can be thought of as a four-letter alphabet. The combination of those four letters make up all of our DNA, and our DNA codes for all the proteins that are ultimately made from DNA. The genetic code is universal. All organisms on this planet use the same genetic code. As far as Romulans, Kardassians, Vulcans, and little green men on Mars–well, we just don't know about their genetic codes. [☞ Gene Expression; Nucleotide; Tissue-Specific Gene Expression]

There are four nucleotides that make up DNA, and there are 20 amino acids that make up proteins. Therefore, it must be true that one nucleotide cannot code for one amino acid. If you consider the possibility that a two-nucleotide combination (call it a twin) coded for each amino acid, you would find that there are only 16 (4^2) possible twins, and that's not enough, either. Through work with the simple genomes and protein architecture of certain viruses, it was found that a string of three nucleotides, called triplets, code for amino acids. If you raise 4 (nucleotides) to the 3rd power, you get 64, and such a coding system can obviously accommodate 20 amino acids.

Work done in the early 1960s by M. Nirenberg, H. Matthaei, S. Ochoa, and P. Leder, which centered on using synthetically prepared nucleotides in various orders and detecting which amino acids were produced, led to the remarkable achievement of the cracking of the genetic code. Several interesting features were deduced. The triplet code has no punctuation: No space or comma occurs between the end of one triplet and the beginning of the next. The start point and reading frame thus become very important, because if something happens that makes decoding proceed from, for example, the normal ABCABCABC to BCABCABCA, then all the amino acids produced from that mutant piece of DNA will be in a garbled, incorrect order. A precise START triplet codon within the genetic code is needed, and in fact such a codon exists. (START refers to: begin protein synthesis from a particular RNA sequence that will be translated into that protein by the cell.) From this it also follows that the introduction or deletion of one or two bases from a coding sequence is much worse than the introduction or deletion of a triplet. In fact, this is true: So-called out of phase or frameshift mutations generally have a worse effect on the resultant mutant protein than mutations that are in phase (insertion or deletion of three nucleotides where only one amino acid is flawed).

Another interesting feature of the code is that it is degenerate. There are more combinations of triplets (64) than there are amino acids (20). In fact, several amino acids are encoded by more than one triplet. For example, alanine is encoded by GCU, GCC, GCA, and GCG; valine also has four different triplets that code for it. Only two amino acids have but a single triplet codon. While degenerate, the code is not imperfect. No triplet codon codes for more than one amino acid.

The first two bases in the triplet codon are more specific (for a given amino acid) than the third base. In the alanine example above, notice that the first two bases are always GC and that any of the other bases found in RNA (U, C, A, or G) can complete the triplet code for alanine. The third base is not as important and tends to "wobble," as Francis Crick, co-discoverer of the double-helical nature of DNA, put it. (Remember, RNA is translated into protein and RNA contains U instead of T, as in DNA. That's why there are U bases in the examples above. Don't forget that this whole process is ultimately determined by the base sequence in the master DNA molecule.)

AUG is the triplet used to signal the initiation of protein synthesis and happens to code for the amino acid, methionine. Three of the 64 triplets code for no amino acid. UAG, UGA, and UAA signal the cellular machinery to end protein synthesis here (at the STOP codon). The protein is done, and so is this entry.

Genetic Counseling

As we learn more about DNA, genetics, the human genome, and the relationships of all of these to human disease, our need for appropriately trained and certified genetics counselors grows. Genetics counselors provide valuable and confidential information to patients when evidence of genetic disease in the family is learned or suggested. Important, sensitive communication must occur so that patients can make intelligent, informed choices that may affect their health, quality of life, reproduction decisions, and more. There are not enough qualified genetics counselors in the United States or the world to deal with the vast amount of genetic data that we are collecting today and will use tomorrow. If you are interested in these fields or want to know more about genetic counseling as a career, contact:

The American Board of Genetics Counselors
☎ 301-571-1825
The National Society of Genetics Counselors
☎ 610-872-7608.

Genetic Engineering

Genetic Engineering as a man-made pursuit is a relatively recent phenomenon. I say that because nature has been engineering new life forms through the manipulation of DNA for eons, through a process called evolution (with apologies to any Creationists reading this). Genetic Engineering as a laboratory phenomenon began with the discovery of restriction endonucleases. We learned how to manipulate DNA *in vitro* (in the test tube) so that we could recombine it (hence the term "recombinant DNA") with other pieces of DNA, insert these so-called clones into bacterial cells, and make lots more as the bacteria (or viruses or yeast) divided. Some also refer to this process as gene splicing, although splicing refers to another, more natural phenomenon. [☞ Clone; Recombinant DNA; Restriction Endonucleases; Gene Splicing]

Genetic Engineering has been refined over the 25 years of its existence to the point where medically and pharmaceutically important reagents, such as insulin and interferon (an antiviral drug), are now routinely manufactured on a large scale. Genetic engineering is also used in brewing, fermenting, wine making, and other fields. Genetic Engineering has contributed significantly to human progress and is a multibillion dollar industry worldwide.

Genome

A master blueprint. The genome of an organism is its complete genetic complement or the complete set of instructions for reproducing that organism and carrying out its biological function in life. The DNA in our cells comprises our genome. When our cells divide, the complete genome in those cells is duplicated for transmission to each of the resultant daughter cells.

The Human Genome Project is a large-scale government undertaking to sequence the three billion base-pairs present in the human genome. [☞ Human Genome Project]

Genotype

Genotype refers to the genetic information contained in an individual organism that is dependent upon the DNA in that individual's genome. Think of it as "gene type." The manifestation of that genotype is known as phenotype. [☞ Phenotype]

Eye color is a good example to explain the difference between genotype and phenotype. Two alleles or forms of a gene are present in each individual. [☞ Allele] A person who has inherited one allele for blue eye color from one parent and one allele for brown eye color from the other parent will have brown eyes. This is because the allele for brown eyes is dominant. This person's genotype is that he or she is heterozygous for brown eye color (a blue allele and a brown allele). The person's phenotype is brown eye color; the actual manifestation of the genotype is brown eyes. Another person who has inherited two brown alleles has the same phenotype–brown eyes–but a different genotype; this person is said to be homozygous for brown eye color.

In terms of disease, let's use as an example a common mutation called ΔF_{508}, which causes cystic fibrosis. ΔF_{508} refers to the deletion of the phenylalanine amino acid normally present at position 508 in a protein, which when mutated, causes cystic fibrosis. Delta (Δ) is for deletion, F is the abbreviation for phenylalanine, and 508 refers to the relevant position in the protein. In the following example, all three individuals have different genotypes. The first two individuals have the same phenotype: neither is affected with cystic fibrosis.

Individual Number	ΔF_{508} Mutation	Genotype	Phenotype
1	Has no mutated alleles.	Normal/Normal	Not a carrier of cystic fibrosis and not affected with cystic fibrosis.*
2	Has one mutated allele.	Normal/ΔF_{508}	Carrier for cystic fibrosis but does not have the disease.
3	Has two mutated alleles.	$\Delta F_{508}/\Delta F_{508}$	Affected with cystic fibrosis.
*at least with respect to this mutation; there are hundreds of other mutations that, if present, are known to cause cystic fibrosis.			

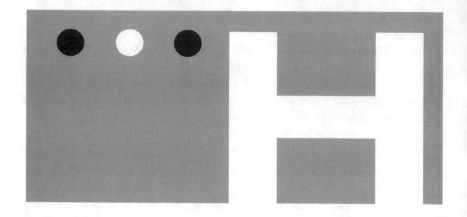

Hairpins

It's hard to believe this word has relevance in a book about DNA–but it does. When designing oligonucleotide primers for use in the laboratory, one needs to be careful to avoid creating a sequence where one end of the oligo happens to share complementarity with the other end of the oligo. If that is the case, the two ends will find each other, base-pair very happily to each other, and form what is known in the field as "hairpins." Good oligonucleotide probes and primers should not form hairpins if they are to be used successfully. [☞ Primers; Probe]

Histone

Histones are proteins. There are five main classes: H1, H2A, H2B, H3, and H4. Histones H2A, H2B, H3, and H4 come together in a specific way, forming what looks like a bead. About 200 base pairs of DNA wrap around this bead and extend, stringlike (on the string, the DNA is associated with histone protein H1) to the next bead. So DNA is organized in the cell like beads on a string. Another term for the bead in this "beads on a string" structure is nucleosome (H2A, H2B, H3, H4, and 200 base pairs of DNA).

Human Genome Project (HGP)

The Human Genome Project is an attempt to sequence all three billion base pairs in the human genome in an effort to learn more about our genetic makeup. The project is an international effort which will take 10 to 15 years to complete at a cost of billions of dollars. It has been highly successful to date and may be completed under budget and ahead of schedule. The goal of the project is to gain insight into disease, aging, and death. Can we reverse or slow these processes? Disease? Probably. Aging? Maybe. Death? Probably not. Taxes? Never!

Hybridization

The process of forming a double-stranded DNA molecule between a probe (created in the laboratory) and a target (patient DNA in the clinical molecular pathology laboratory). DNA is double-stranded and can be made single-stranded. If the two strands find each other again, they are said to have reassociated with each other. If however, the investigator adds a large excess of DNA, called a probe, which is complementary to a particular sequence of interest, the probe, just based on competition and numbers, finds the target before the sister strand does. That process is called hybridization because a hybrid duplex (target to probe) has been formed instead of simple reassociation of the two sister strands. [☞ Complementary Strands of DNA; Duplex; Probe]

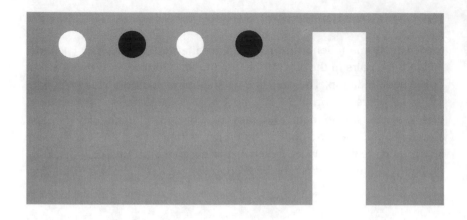

Identity Testing

[☞ Paternity/Profiling/Identity/Forensic Testing by DNA]

in vitro, in vivo, in utero

Experiments or work done in a test tube or some other kind of laboratory container are said to be done *in vitro*. DNA mutations may be artificially induced *in vitro*. Experimental or therapeutic work done inside the body is said to be done *in vivo*. DNA mutations may be induced in experimental laboratory animals *in vivo* or by environmental insult in people *in vivo*. *In utero* means in the uterus.

Intron

The opposite of "exon." Introns are intervening stretches of DNA that separate exons. They are spliced out of the gene at the RNA level and are not ultimately expressed as part of the gene product (protein). [☞ also Exon; Junk DNA]

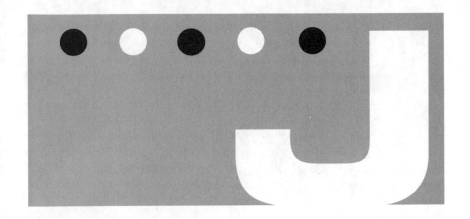

Junk DNA

An artifact of DNA replication. The enzyme that makes new DNA (DNA polymerase) can't make up its mind and goes back and forth between different points in the replicating DNA molecule that it recognizes as replication start points. Lots of extraneous DNA, called junk DNA, is made. When viewed under the electron microscope, this complex of replicating and aborted replicating DNA looks something like a crazy roadmap. [☞ also Polymerase; Replication]

Some scientists have called introns "junk DNA." This is probably an oversimplification, as introns serve some sort of regulatory or punctuation role within genes and are not strictly junk. But then, even scientists need to label things so they can put them in a convenient little compartment in their brains reserved for stuff they don't want to think about very much. [☞ also Exon; Intron]

kb

Kilobase; 1000 bases. A gene that is 4000 base pairs long is said to be 4 kb (or kbp for kilobase pairs; remember, DNA is double-stranded) in length.

Kinase

Kinases are a class of proteins that add phosphorus molecules to their substrates. (A substrate is the molecule that the business end of the protein deals with.) Phosphorus can be made radioactive and used to "label" DNA for use in the DNA diagnostics laboratory. If we have successfully used a radioactive DNA probe to find another (complementary) piece of DNA in a patient specimen or from DNA extracted from a bloody glove, then we can detect that radioactive hybrid (the probe DNA and the target DNA find each other and form a hybrid) visually on an X-ray film.

Labeling

[☞ DNA Labeling; Autoradiogram; Chemiluminescence]

Lagging Strand

The academic laggard portion of the DNA double helix. As it pertains to DNA replication, the lagging strand refers to a specific thing. Growth is dependent on cell division (1 cell becomes 2, 2 become 4, etc., etc., etc.), and cell division is dependent on DNA replication. DNA is a double helix, composed of two strands. Think of the two strands as sister strands. When the cell reaches that portion of the cell cycle where it's time to divide, the two sister strands separate slightly in the middle, forming a so-called replication bubble. The molecules and enzymes involved in the business of DNA replication make more DNA bidirectionally. What that means is that a new daughter strand is made off of one sister strand in one direction and a second daughter strand is made off of the other sister strand in the other direction. The daughter strand synthesized in the right-to-left direction is the leading strand. The daughter strand synthesized in the left-to-right direction is the lagging strand.

Lambda (λ) DNA

Lambda (λ) phage is a virus that infects bacteria. It has a relatively small genome close to 50,000 base pairs (50 kilobases or kb) in length. Its genome can be cut into fragments of known size by different restriction endonucleases. When these fragments are electrophoresed, they migrate to a particular position based on their molecular weights. Since we know the sizes of those fragments, we can use them as molecular weight "standards" against which we can "size" DNA fragments whose molecular weights are unknown. This is actually common practice in the clinical molecular pathology laboratory for purposes of quality control of the laboratory test results generated. [☞ Bacteriophage; Electrophoresis]

Leading Strand

The academically gifted portion of the DNA double helix. As is pertains to DNA replication, the leading strand refers to a specific thing. For an explanation, ☞ Lagging Strand.

Ligase Chain Reaction (LCR)

LCR is a PCR "wannabe." [☞ PCR "Wannabes"] LCR is an *in vitro* nucleic acid amplification technique developed and marketed by Abbott Laboratories in Abbott Park, Illinois. In 1996, LCR-based tests for detection of the bacteria *Chlamydia trachomatis* and *Neisseria gonorrhoeae* were approved for diagnostic use by the U.S. Food and Drug Administration.

LCR may potentially rival PCR as an important diagnostic tool for the clinical laboratory, due to its sensitivity, speed, and adaptability to automation. LCR depends upon mixing target (patient) DNA, a thermostable enzyme called DNA ligase, oligonucleotide DNA probes, and other ingredients. The mixture is heated to denature all double-stranded DNA present (target DNA and complementary probes). Two pairs of complementary probes are used that have complementarity to the nucleotide sequence of the DNA target of interest. After heat denaturation and subsequent cooling, the four probes bind to their complementary sequences on each sister strand of the target DNA. The two probes that bind to each strand are designed such that when they bind a small gap exists between them. The DNA ligase present ligates (joins) the two bound probes together, thereby achieving a "doubling" of target DNA. As this process continues through additional cycles, an exponential amplification of the amount of target DNA occurs because the ligated molecules (called amplicons) also serve as targets for probes. A sensitive detection technique reports a positive result if the bacterial target was present in the patient specimen and a negative result if the target was absent.

Locus

Remember that Old Testament stuff about Moses, the Israelites, and their slavery in Egypt? Well, as it happened ten plagues were brought on the Egyptians in order to "convince" the Pharaoh to "Let my people go." The eighth plague, locusts, was particularly gross. Oh—wait a minute, that's locusts, with a "t." Sorry. . . .

Locus is a fancy word for place. The locus for a particular gene in the human genome is the place where you would find that gene. Jargon has also turned the word "locus" into a synonym for group, as in the HLA locus of genes or the XYZ locus.

Major Groove

When things are going just great, you're in a major groove. With respect to DNA, though, the major groove is the larger of the two indentations repeated in a regular fashion throughout the DNA double helix. The DNA double helix makes a complete turn and begins a new one every ten bases. The way the two strands of nucleotides that form the double helical structure of DNA wind around each other causes grooves within the molecule, one relatively small indentation or groove and one larger one. The small one is the minor groove and the larger one is the major groove. [☞ Minor Groove]

Mendel, Gregor

An Augustinian monk who worked out many of the concepts of heredity and heritable traits using pea plants that he pollinated and bred at the monastery. In 1865, Mendel published his work, "Experiments on Hybrid Plants," in the *Proceedings of the Natural History Society of Brno*. Mendel's is considered classic work.

Messenger RNA (mRNA)

DNA is transcribed into mRNA, which is ultimately translated into protein. [☞ Anticodon]

Miescher, Frederick

The Swiss scientist who discovered DNA in 1869. Miescher used as starting material (OK, here's fair warning–you may not want to go on if you've eaten recently or are considering eating soon) pus and salmon sperm (really!) and named the stuff that he isolated from the nuclei of cells "nuclein," which was later identified as DNA.

Minor Groove

When things are going pretty well but not as great as when you're in a major groove. In DNA, the minor groove Is the smaller of the two indentations repeated in a regular fashion throughout the DNA double helix. It occurs to me that a figure would be really useful here, but if I started putting in figures to illustrate every point, then we would've had to charge more for the book, now wouldn't we? Actually, you can appreciate the grooves by examining the computer graphics of Paul Thiessen that appear on the cover of this book. [☞ also Minor Groove]

Mismatch

Several formerly married couples that I know. With respect to DNA though, mismatch refers to lack of complementarity. DNA is made up of nucleotides called adenine (A), cytosine (C), thymine (T), and guanine (G). The laws of base pairing say that, in DNA, A always base pairs to T and G always base pairs to C; and so goes the double helix. Usually, though, when you see the word "always," you know the author's about to tell you about the exceptions. Well, yes, mistakes happen, mutations occur, and a G can find itself across from a T or a C finds itself mismatched with not a G, but (heavens) an A. That's mismatching. It's to be avoided and our bodies have evolved a biochemical way to minimize its occurrence. [☞ also Complementary Strands of DNA]

Mitochondrial DNA (mtDNA)

Mitochondria are distinct elements within animal and human cells. Such a distinct element or subunit of the cell is called an organelle. (The nucleus is another example of an organelle.) Mitochondria are involved in oxygen transfer and energy conversion. (You may remember learning in elementary school that mitochondria were the "powerhouse of the cell.") Mitochondria contain their own DNA distinct from the nucleus, called mitochondrial DNA, or mtDNA. mtDNA is derived only from one's mother. The mitochondrial genome is small compared to the nuclear genome, at about 16,500 bases, and it has no introns. [☞ Intron] In man, mtDNA has 13 protein-coding regions.

The mutation rate of mtDNA is greater than that of nuclear DNA. Mutations in mtDNA are known to cause human disease, particularly in the brain, heart, liver, kidney, muscle, and pancreas. Examples include Leber's hereditary optical neuropathy, Pearson syndrome (bone marrow and pancreatic failure), and myoclonic epilepsy and ragged red fibers.

Molecular Biology

Molecular biology is the field of endeavor in which I received my Doctoral degree. Molecular biology is the study of the business of life at the level of the lowest common denominator: the molecules that carry it (life) out–DNA, RNA, and proteins. Molecular biology as applied in the clinical diagnostics laboratory has been given the moniker of molecular pathology. Applications to genetics are called molecular genetics.

Molecular Pathology

Pathology is the study of those stimuli that cause disease and the examination of tissue affected by disease. Molecular pathology is the application of the tools of molecular biology (DNA technology) to the medical practice of diagnostic pathology.

mRNA

[☞ Messenger RNA; Ribonucleic Acid (RNA)]

mtDNA

[☞ Mitochondrial DNA]

Mutation

A mistake. A mutation in DNA is an error or permanent alteration that has occurred in the coding sequence of a gene or genetic regulatory element. Such an error occurs during replication of the DNA and may be a result of environmental insult that has somehow disturbed the DNA sequence, or it may be due to an error introduced naturally by the DNA copying machinery of the cell. Sometimes advantageous mutations have been introduced "on purpose" by nature in an effort to deal with a particular problem, such as too many malaria-causing insects flying around or not enough sunlight because of a natural or man-made disaster. Mutations help species evolve as necessary and are the biological mechanism behind Charles Darwin's theory of "survival of the fittest." (Again, I find myself apologizing to Creationists reading this; maybe they're wrong; maybe I'm wrong–God only knows.)

Sickle cell anemia is caused by two mutations in a particular gene and is a serious clinical condition. However, children with the mutation that causes sickle cell trait have natural resistance to a fatal form of malaria in Africa. Individuals who have one normal copy of the gene and one mutated copy of the gene (sickle cell anemia) live fairly normal lives.

Of course, mutations are not necessarily a good thing. Mutations can have a range of effects on organisms, from no effect (silent mutation) to carrier status to deleterious and damaging effects that lead to outright disease or death. Some mutations that occur *in utero* are incompatible with life. There are several classes of mutations: nonsense; missense; frameshift; insertion, deletion, and point mutations.

Nonsense: Erroneously introduced into the reading frame of a gene is a codon for STOP which causes growth of the protein chain to terminate prematurely. The protein is shorter than normal and may be partially functional, largely functional, or totally non-functional. β-Thalassemia, a chronic anemia, is caused by many different kinds of mutations, some of which are nonsense mutations in the gene for β-globin.

Missense: The DNA coding sequence for the gene has had one triplet codon altered such that a different amino acid is substituted for the one that is typically present. Some mutations in the Low-Density Lipoprotein Receptor Gene are missense mutations and cause Familial Hypercholesteremia, leading to coronary heart disease.

Frameshift: The introduction of some number of bases, not divisible by 3, into the reading frame of a gene. Codons in a gene are composed of 3 bases, and if a base is added or deleted, the order of the amino acids encoded for by that gene is wrecked. The result may be a prematurely truncated protein (if a STOP codon is created where there wasn't one before), or a protein that has little relationship at the amino acid level to the normal protein because it is composed of a very different amino acid sequence. The result is usually a bad one. In the 185delAG mutation, for example, two bases, adenine and guanine, are deleted from exon 2 of *BRCA1,* altering the translational reading frame of the subsequent mRNA. The frequency of 185delAG in Ashkenazi Jewish women is 1 in 107 and the mutation is associated with the onset of breast cancer in this group before the age of 40. This mutation is a candidate for screening in this population.

Insertion: One or more bases is inserted. A 1 or 2 (or any number not divisible by 3) base pair insertion is a kind of frameshift mutation.

Deletion: Bases are deleted. A deletion can also cause a frameshift mutation if the number of bases deleted is a number not divisible by 3. The ΔF_{508} mutation that causes cystic fibrosis is a deletion mutation.

Point: A point mutation is one in which a single base pair has been changed; it can be a substitution, an insertion or a deletion. Substitutions that don't change the amino acid that is encoded by the triplet codon are silent mutations. For example, GCC codes for the amino acid alanine but so does GCA, so if that mutation occurs (C mutated to A at position 3) there's no effect on the gene product (the protein). Some point mutations are deleterious, such as the single base change that results in sickle cell anemia.

Splice Site Mutation: ☞ Gene Splicing

NASBA

[☞ Nucleic Acid Sequence-Based Analysis]

Nick Translation

Nick translation is a commonly used biochemical procedure in the molecular biology laboratory. It is an enzymatic method of labeling DNA probes with radioactively or otherwise tagged deoxyribonucleotides that are incorporated into newly made DNA molecules during the course of the reaction. Once tagged, these DNA probes can be used as reporter molecules in subsequent tests to answer questions about DNA or RNA obtained from a patient specimen. [☞ also Autoradiogram; Chemiluminescence; DNA Labeling; Oligonucleotide Priming; Southern Blot]

Northern Blot

The northern blot technique is essentially the same thing as the Southern blot. [☞ Southern Blot for a more intensive explanation.] The key difference is that patient DNA is the target of investigation in Southern blotting, while RNA is the target in the northern blot. Unlike Southern blots, northern blots are not routinely used in the clinical molecular pathology laboratory.

While the Southern blot was developed by a person, Dr. E. Southern, there was no "Dr. Northern" (which is why one capitalizes Southern
but not northern blot). The northern blot derives its name from the fact
that it is, in a way, the opposite of the Southern blot with respect to the
target of investigation.

Nuclease (DNase and RNase)

A nuclease is a protein whose job is to digest nucleic acids or nucleotides. A nuclease can be RNA-specific (an RNase) or DNA-specific
(a DNase). Action can be internal to the nucleic acid (endonuclease,
"endo" meaning inside or within) or from the end where the nuclease
"chews off" one nucleotide at a time (exonuclease, "exo" meaning outside or at the end). These enzymes have been co-opted by molecular biologists for use as tools in the molecular biology laboratory. Restriction
endonucleases are a type of nuclease used all the time. DNase I has a
place in nick translation. [☞ Nick Translation] There are numerous
other examples.

The normal role of nucleases in the body is to break down, enzy-
matically, any nucleic acids ingested during eating. The pancreas is an
organ rich in nucleases. Degradation of nucleic acids occurs in the intes-
tine using the nucleases secreted by the pancreas. Rattlesnake and
Russel's viper venom also contain nucleases that are not particular;
these enzymes work equally well to degrade DNA or RNA.

DNA and RNA purification for molecular pathology investigation of
tissues like the pancreas or nuclease-rich tumors must proceed quickly
to inactivate nucleases before they can act on the DNA and RNA present.

Nucleic Acid

A class of naturally occurring biochemical entities. DNA and RNA
are the two prime examples. Nucleic acids are composed of sugar molecules, nitrogenous bases, and phosphate groups; when one of each of
these joins, a nucleotide is formed. When nucleotides become chemically joined to each other, nucleic acids are formed. If the sugar mole-

cule is a ribose containing sugar, the nucleic acid formed is ribonucleic acid (RNA). If the sugar molecule is a deoxyribose (missing one oxygen molecule) containing sugar, the nucleic acid formed is deoxyribonucleic acid (DNA).

Nucleic Acid Sequence-Based Amplification (NASBA)®

NASBA is an *in vitro* nucleic acid amplification technique, a PCR "wannabe." [☞ PCR "Wannabes"] NASBA is a registered trademark of Cangene Corporation of suburban Toronto and is being developed and distributed by Organon Teknika. Unlike Polymerase Chain Reaction (PCR), which accomplishes amplification of input DNA by cycling among different temperatures to accomplish the necessary biochemistry, NASBA is isothermal: all phases of the biochemistry occur at a single temperature (approximately 40° C). NASBA accomplishes nucleic acid amplification using three enzymes: Reverse transcriptase, RNase H, and T7 RNA polymerase. [☞ Reverse Transcriptase] Two primers are also used to initiate the different reactions that occur. [☞ Primers]

NASBA has been used to detect the presence of RNA containing viruses like Hepatitis C Virus and Human Immunodeficiency Virus. Reverse transcriptase forms DNA from viral RNA, if present in the patient specimen, using primer number 1. The RNA in the resultant RNA:DNA hybrid is destroyed by RNase H, an enzyme that specifically chews up RNA in such hybrids (hence the "H"). The remaining RNA participates in further reactions using reverse transcriptase and T7 RNA polymerase (a bacterial virus called T7 is the source of this RNA polymerase) and the second primer to generate an exponential amplification of the RNA in about 90 minutes. Typical amplification is on the order of a billion-fold increase in the amount of input RNA.

Two of the primary advantages of NASBA over PCR are that one can use RNA without first having to turn it into cDNA; and that the NASBA reaction may proceed at one temperature, eliminating the need for expensive thermal cycling devices; one can use an ordinary water bath. [☞ cDNA; Polymerase Chain Reaction (PCR)]

Nucleoside

[☞ Nucleotide]

Nucleosome

DNA wound around histones forms a structure known as a nucleosome, important in the compression of DNA so that the very long DNA strands present in a nucleus physically fit there. [☞ Histone]

Nucleotide

In the same way that amino acids are the building blocks of proteins, nucleotides are the building blocks of the nucleic acids, DNA and RNA. Nucleotides are composed of phosphate groups, a five-sided sugar molecule (ribose sugars in RNA, deoxyribose sugars in DNA), and nitrogen-containing bases. These bases fall into two classes: pyrimidines and purines. Pyrimidines are chemically distinct from purines and include cytosine (C), thymine (T), and uracil (U). (Uracil is usually found only in RNA.) Purines include adenine (A) and guanine (G). A nucleotide without its phosphate group is called a nucleoside.

Nucleotide abbreviations are as follows:

The first letter indicates the base: A, C, G, T or U.
The second letter indicates whether 1, 2, or 3 phosphate
   groups are present: M for mono, D for di, or T for tri.
The third letter is always P, for phosphate.

If this three-letter abbreviation is preceded by a lower case "d," that is a designation for a deoxyribonucleotide, a DNA building block. If there is no "d" prefix, the nucleotide is understood to be a ribonucleotide, an RNA building block. Examples are: AMP for adenosine monophosphate and dTTP for deoxythymidine triphosphate.

OJ

A commonly used abbreviation for orange juice.

Oligonucleotide ("Oligos" is the slang term)

If you string together a few nucleotides you have an oligonucleotide or an oligo (oligo means few). This artificial synthesis can be done in very expensive machines that you can buy and put in your laboratory (or garage if you're so inclined I suppose). Oligos are absolutely essential as primers for the polymerase chain reaction (PCR) and therefore for clinical molecular pathology. We usually take advantage of the oligo price war that is going on by calling a supplier's toll-free number, telling them the sequence of the oligo we need, and then presto, we get it by courier the next day for about $1–1.50 a base. [☞ PCR; Primers]

Oligonucleotide Arrays

[☞ DNA Chips]

Oligonucleotide Priming

Oligonucleotide priming is a commonly used biochemical procedure in the molecular biology laboratory. It is an enzymatic method of labeling DNA probes with radioactively or otherwise tagged deoxyribonucleotides that are incorporated into newly made DNA molecules during the course of the reaction. Once tagged, these DNA probes can be used as reporter molecules in subsequent tests to answer questions about DNA or RNA obtained from a patient specimen.

The word "priming" is used because the reaction depends upon DNA polymerase making new DNA strands off of a template. Such new DNA synthesis only begins if the DNA polymerase has a local region of double-strandedness to initiate DNA synthesis. The double-strandedness occurs through the addition to the reaction of short oligonucleotides (6 to 10 base pairs in length). The sequence of these oligos is random; they find many complementary areas in the DNA to be labeled and bind there. Then DNA polymerase in the reaction can go about its business of making new DNA, incorporating into the newly made DNA strands tagged or labeled deoxyribonucleotides, also present in the reaction mixture. The tag or label may be radioactive or some other chemical entity that will allow detection of hybrids formed between the DNA probe and its targets later on in the laboratory test.

Since the sequence of the oligos is random, some refer to this biochemical reaction as random oligonucleotide priming. Originally, when the oligos used were six base pairs in length, this technique was known as random hexanucleotide priming. [☞ also Autoradiogram; Chemiluminescence; DNA Labeling; Nick Translation; Oligonucleotide; Primers; Southern Blot]

Oncogene

A cancer-causing gene. We have many genes involved in controlling cell division and the rate of cellular growth. These genes have a normal, useful function and are called proto-oncogenes. When proto-oncogenes

61

are mutated, through any one of a number of mechanisms (throw away those cigarettes), they lose their ability to regulate cell growth and become cancer-causing oncogenes. Examples of oncogenes include *abl*, *erbB*, *ras*, and *myc*. These and other oncogenes have been implicated in breast cancer, colon cancer, neuroblastoma (a kind of brain tumor), various kinds of leukemia and lymphoma, and other cancers.

Open Reading Frame (ORF)

ORF is perhaps the loneliest acronym in the field of molecular pathology. In the nucleotide sequence that comprises a gene are stretches of bases that will ultimately be translated into a protein. Each three successive bases, termed a triplet, codes for a corresponding amino acid. (Amino acids are the building blocks of proteins.) There are three triplets that code for STOP signs: when the cellular machinery involved in elongating the protein chain that is being translated from the DNA and RNA that code for it encounters a triplet that signals STOP, the protein growth terminates and a mature protein (or prematurely terminated, mutant protein) has been generated. An ORF, then, is a stretch of bases in DNA that *could* code for a protein because it has a specific START triplet and no STOP triplets (at least for a while, until a reasonably sized protein can be generated from that stretch of bases). [☞ also Genetic Code]

Paternity/Profiling/Identity/Forensic Testing by DNA

You know those bar codes that cashiers in supermarkets scan to figure out whether you're buying a 99-cent can of peas or a $14 bottle of wine? This is somewhat analogous. In the same way you individualize the peas or the wine, one can use DNA patterns to individualize DNA specimens. Reasons for wanting to individualize DNA specimens are to determine paternity/non-paternity, or to search for matches between suspects and biological samples left at a crime scene.

All individuals can be distinguished from each other at the DNA level. This is because the DNA of each individual is different, at several different levels, from all other individuals (except for identical twins, whose DNA sequences and patterns are identical). These so-called "genetic signatures" (the bar codes) can be identified in the laboratory by Southern blot and Polymerase Chain Reaction based assays which exploit DNA polymorphisms (a fancy word for "difference").

In general, polymorphisms refer to different forms of the same basic structure. At the DNA level, polymorphisms are evident in different ways. The most significant one for identity testing is the different number of repeats in a repetitive DNA sequence. For example, AGCT may be repeated 62 times in tandem in one person and 47 times in another person, and that difference is detectable. Repeated sequences in DNA have been termed minisatellite DNA, or Variable Number Tandem Repeats (VNTRs). The number of repeat units within minisatellite DNA is highly variable, both within a single genetic locus and among different genetic loci. Different probes for several core sequences that comprise different hypervariable regions exist. When DNA is hybridized with a probe specific for multi-locus hypervariable sequences, a complex pattern of bands (which looks much like a supermarket bar code) appears on an autoradiogram, and this pattern is unique for every individual. Alternatively, one may use probes specific for single locus hypervariable regions that are highly polymorphic–that is, highly variable from person to person. The probability of two individuals having the same number of alleles in these highly polymorphic regions is quite low. Individualizing power becomes very great in this mode of analysis when additional single locus probes are used.

Minisatellite repeats are also termed VNTRs. Within our genomes we have segments of DNA that are variable in number (the VN in VNTR) and that reiterate a particular identical sequence within that segment of DNA, over and over, a so-called tandem repeat (the TR in VNTR). The number of repeat units within a minisatellite repeat or VNTR is highly variable among individuals and can be determined, as described above, for purposes of identification.

There are several PCR-based methods that may be used to identify human DNA polymorphisms for purposes of individualizing and identifying them. Analysis for several different genes and genetic loci that exhibit polymorphisms among individuals include the HLA DQα locus, low-density lipoprotein receptor (LDLR), glycophorin A (GYPA), hemoglobin G gammaglobulin (HBGG), D7S8, and group-specific component (GC). Determination of "length polymorphisms" by PCR of amplified fragment-length polymorphisms (AMP-FLPs) present in variable number tandem repeats (VNTRs) in the human genome may also be done.

Half of an individual's DNA is inherited from each biological parent. Therefore, the DNA testing (also known as "DNA fingerprinting") described can be used to include or exclude an alleged father from that group of men that could be the biological father of a child. Similarly, it could be used to establish maternity. Immigration questions also sometimes hinge on paternity and sometimes involve DNA fingerprinting.

There are, of course, applications of identity testing to forensic and criminal investigation. An individual suspected of having committed a crime can be placed at the scene of the crime if the suspect's DNA fingerprint matches the fingerprint obtained from DNA of hair, blood, semen, etc., left at the crime scene. This would be valuable evidence for the prosecution of that suspect. At the same time, a non-match between the DNA of the evidentiary material and the suspect can help exonerate the individual. DNA fingerprinting was first applied in two cases of rape/murder in the mid 1980s by Sir Alec Jeffreys in the United Kingdom.

There are also clinical applications of DNA fingerprinting. Bone-marrow donors and recipients may be "fingerprinted" before a bone-marrow transplant is done. Such information may be used post-transplant to determine the success or failure of the engraftment procedure by analyzing the identity of the cells in the bone marrow of the recipient. DNA fingerprinting may be used for genetic investigation to determine whether twins, triplets, etc. are monozygotic (identical), or dizygotic (fraternal). Prenatal testing for genetic disease is often preceded by or done simultaneously with analysis of the gene in question of the parents of the unborn fetus. Obviously, it is crucial that analysis be done on the biological parents, so DNA fingerprinting may be an important adjunct test.

PCR (Polymerase Chain Reaction)

You've heard of looking for a needle in a haystack; PCR is a biochemical reaction that generates a haystack full of needles in a very specific way. It's a biochemical procedure for *in vitro* nucleic acid amplification. Here's how it works:

Patient DNA is purified in a reaction tube that contains all the ingredients necessary for PCR and heated to a temperature near boiling (usually around 94–95° C). This high temperature denatures the DNA strands–that is, makes the naturally double-stranded DNA single-stranded. This is called the *denaturation* step. Present in the reaction mixture are oligonucleotide primers. I'll get back to these primers and how they are used to make PCR so specific. The temperature of the reaction is cooled to anywhere from 30–65° C, and the primers find the bases in the patient DNA to which they are complementary and bind there. This is called the *annealing* step. This creates a local region of double-strandedness and the DNA polymerase present makes use of the building blocks of DNA, deoxyribonucleotide triphosphates (dNTPs for short), to make more DNA using the patient DNA as a template for synthesis. This occurs when the temperature of the reaction is changed to 65–75° C and is called the *extension* step. This cycle of denaturation, annealing, extension is repeated 25–40 times.

Each cycle accomplishes a doubling of the amount of DNA that was present before. Add 1 double-stranded DNA molecule, denature it into 2 single strands, anneal the primers to each strand, extend those primers to make new DNA, and you've got 2 double-stranded DNA molecules where you only had 1 before. Enter the second cycle and those 2 double-stranded DNA molecules create 4, then 8, then 16, then 32, and so on. After about 30 cycles a billion-fold increase in the amount of starting DNA has been accomplished. Now, instead of a needle in a haystack you've made a haystack full of needles by specifically copying them.

The clever among you will realize that proteins, like the DNA polymerase necessary in PCR, don't do well at temperatures as high as 94° C. In fact, most proteins become unraveled and destroyed and do not function at such high temperatures. An important technical advance for PCR and one that led to its automation came with the realization that there are bacteria that normally carry on the business of life in hot springs (like the ones in Yellowstone National Park). The bacterium, *Thermus aquaticus*, lives in such springs at temperatures of 75° C and the DNA polymerase purified from this bacterium functions at temperatures over 90° C. This DNA polymerase, termed *Taq polymerase* after the bacteria from which it is purified, is the workhorse of PCR and has been a significant factor in the wide use of PCR in the clinical molecular pathology laboratory.

The specificity of PCR is dictated by the sequence of the primers used in the reaction. We need to know the sequence of the DNA fragment of interest. Choose primers that bind to the 2 ends of the sequence of interest which is, for example, 250 base pairs in length (sequences of 200–500 base pairs work well). The primers need to be long enough

(15–25 bases) so that they bind only to the specific regions of interest and not randomly throughout the genome. Once this specific primer binding occurs, PCR works efficiently to create more and more of that 250-base-pair-long piece of DNA so that we can analyze it in the laboratory.

Objects of PCR investigation are not limited to searching for a particular genetic mutation. In other words, PCR is not limited to examining only the patient's DNA. If an infectious disease is suspected, PCR can be used to detect the presence of a piece of DNA that is specific for the microorganism in question. That DNA will have come along for the ride during the purification of the DNA from that particular clinical specimen, such as a blood specimen. If detected, then the PCR test for that particular bacteria or virus is positive.

Some viruses are RNA viruses by nature, for example, Hepatitis C Virus and Human Immunodeficiency Virus. RNA is not a suitable starting material for PCR. But one extra step solves that problem. Reverse transcriptase (RT) is an enzyme that naturally synthesizes DNA from RNA as starting material. RNA plus RT yields DNA that is designated cDNA ("c" for complementary) to denote this fact. cDNA is then a DNA molecule that is perfectly suitable for subsequent PCR. Think of it as RNA PCR; it is abbreviated RT PCR, for Reverse Transcriptase Polymerase Chain Reaction. An enzyme called *Tth* polymerase has the ability to combine the activities of reverse transcription and the important enzyme in PCR, DNA polymerase, whose job is to make more DNA. *Tth* polymerase is a thermostable (stable at high temperatures) enzyme from the bacteria, *Thermus thermophilus;* hence the name.

The biochemistry of PCR was developed in the mind of Dr. Kary Mullis on an evening drive through Northern California Redwood country. He made it work in the laboratories of Cetus Corporation where he was employed. He didn't have *Taq polymerase,* of course, so he had to add new DNA polymerase with every cycle because the near-boiling temperatures needed for denaturation irreversibly denatured the DNA polymerase. Today we have thermal cyclers that change the temperatures necessary for PCR to work in a rapid, automated fashion. Early PCR, including the work of Dr. Mullis, depended on dedicated scientists sitting by the tubes involved with PCR with a stopwatch and several water baths, set to different temperatures. When the time at one temperature was up, the tubes would have to be placed manually into the next water bath, and so on, and so on, and so on.

Throughout the late 1980s and early 1990s, PCR took the scientific community by storm and is probably the single most important discovery that led to the field of molecular pathology, keeping in mind of course that Southern blotting, nucleic acid extraction, restriction endonuclease digestion, etc., are also very important. It is PCR, however, that is largely responsible for the genetic revolution we are witnessing today. The rights to this biochemical reaction were purchased by Hoffman-LaRoche for $300,000,000, which, as you might imagine when such large sums of money are involved, precipitated litigation. (Speaking of litigation, I have heard Dr. Mullis advise that if you ever have a choice between putting your name next to "Patent Inventor" or "Patent Assignee," choose the latter; it is much more lucrative.) PCR is covered by patents owned by Hoffman-Laroche. [☞ also cDNA; Denature; Oligonucleotide; Primers; Retroviruses; Reverse Transcriptase

"PCR in a Pouch"

This is a variation on the PCR theme. This modification of the way PCR is performed is an exciting technological development, at least to the people who work in clinical diagnostic molecular pathology laboratories. (It doesn't take a lot to get us excited.) All reagents necessary for PCR are placed in plastic bubbles or blisters inside a rectangular plastic pouch. DNA is prepared from a patient specimen and then inserted into a pouch which is then sealed. The pouch is placed inside an instrument which has rolling devices that move the prepared DNA from blister to blister where different parts of the biochemistry involved in PCR and detection of the resultant products take place. The instrument also acts as the thermal cycler necessary for successful PCR to occur. Detection for the presence or absence of an infecting organism's DNA or a particular genetic mutation is also done automatically. The results are displayed as colored dots. The instrument reads the colored dots and then prints the results.

So far, PCR is the closest thing to the "black box" molecular pathologists have always wanted: an instrument in which the specimen goes in one end and the answer comes out on the other, during which time we can walk away from the machine. "PCR in a pouch" is a product of Johnson and Johnson Clinical Diagnostics in Rochester, New York. It should be in wide use in Europe and North America by the end of the 1990s.

PCR "Wannabes"

It didn't take long to become clear that PCR was a powerful technology. Not only did it have tremendous research applications, but people realized it was going to revolutionize medical diagnostic testing. Today, several PCR tests have been FDA-approved for diagnostic use, and many more are somewhere in the pipeline. This situation is another example of how medicine, law, and money (i.e., markets) intersect. PCR is medically important, has a tremendous potential market, and is covered by patents owned by Hoffman-LaRoche. Realizing this, diagnostics companies set their research scientists to the task of developing biochemical reactions similar to PCR: PCR "wannabes" as I call them. PCR "wannabes" have been developed by many different companies, and they are quite good. Some involve complicated biochemistry while others use more straightforward biochemistry, but the end result is the same: *in vitro* nucleic acid amplification (making a haystack full of specific needles in the test tube). Some of those that are already to market include:

Ligase Chain Reaction (LCR), marketed by Abbott Laboratories in Abbot Park, Illinois

Transcription Mediated Amplification (TMA), marketed by Gen-Probe, Inc. in San Diego, California

Branched DNA amplification, marketed by Chiron Corporation in Emeryville, California [☞ bDNA]

Nucleic Acid Sequence Based Amplification, marketed by Organon-Teknika, which has facilities in Europe and North Carolina [☞ Nucleic Acid Sequence-Based Amplification (NASBA)]

Phenotype

The manifestation of a particular genotype of an organism. Examples of phenotypes include blue eye color, affected with cystic fibrosis, and maleness. [☞ Allelle; Genotype for more discussion.]

Plasmid

A plasmid is a circular piece of DNA that exists outside and separate from the chromosome of a bacterial cell. Plasmids are smaller than the bacterial chromosome and many replicate autonomously–that is, independently of the rest of the DNA in the bacterial cell. Molecular biologists have learned how to use plasmids for cloning. Plasmids are commonly used cloning vectors. A piece of DNA of interest is isolated and inserted into a plasmid, and this recombinant piece of DNA is then introduced back into a bacteria where the plasmid will thrive and grow. As it replicates, more and more of the inserted DNA is also made. This is an example of genetic engineering. [☞ also Clone; Genetic Engineering; Yeast Artificial Chromosome (YAC)]

Molecular pathologists exploit the plasmid present in the bacteria *Chlamydia trachomatis*. This bacteria causes the most common sexually transmitted disease. There are 3–4 million cases per year in the United States, and approximately 3 million cases in Europe. Untreated *C. trachomatis* infection has important health consequences, including pelvic inflammatory disease and infertility, among others. There exists a Polymerase Chain Reaction based and a Ligase Chain Reaction based test for the detection of this microorganism in patient specimens (cervical swabs, urethral swabs, and urine) that amplify *C. trachomatis* specific DNA sequences. The target in these tests is DNA in a bacterial plasmid that is present at about 10 copies per bacterial cell. By using this as the target, the "deck is stacked" in favor of sensitive detection because the target has been naturally amplified by the bacteria. The plasmid DNA is therefore a better target of detection than a portion of the bacterial chromosome that is present at only 1 copy per bacterial cell.

Polymerase

An enzyme (protein) whose job is to polymerize, meaning "make more." DNA polymerase is the enzyme involved in making more DNA; RNA polymerase works to synthesize RNA. By the way, these proteins are also coded for by their genes within the DNA of the organism, whether that be a human, animal, plant, bacterium, or virus. [☞ also Gene Expression]

There are several classes of polymerases. There are those that make DNA or RNA, as described above, but this subdivision goes further. A nucleic acid polymerase acts using another nucleic acid as the template to direct that synthesis. The template can be DNA or RNA, depending on the enzyme. There are four general kinds of polymerases, then:

DDDP	DNA-Dependent DNA Polymerase
DDRP	DNA-Dependent RNA Polymerase
RDDP	RNA-Dependent DNA Polymerase*
RDRP	RNA-Dependent RNA Polymerase

*also known as Reverse Transcriptase [☞ Reverse Transcriptase]

Polymerase Chain Reaction

[☞ PCR]

Primers

Think about painting your room or your house. There are some surfaces that paint won't stick to very well so you first cover the surface with a primer, then you can apply the paint successfully and stand back and admire your excellent handiwork. This is a perfectly analogous situation to the biochemistry of primers as they relate to DNA molecules. An enzyme (protein) called DNA polymerase (paint brush) needs raw materials (paint) to make more DNA. DNA polymerase makes more

DNA using unwound DNA as a template. This unwound DNA is single-stranded, so if that strand reads ATTAGCC, it directs the synthesis of a new strand complementary to it: TAATCGG. [☞ Complementary Strands of DNA] But DNA polymerase won't work without a small section of double-stranded DNA to initiate or "prime" new DNA synthesis. In Polymerase Chain Reaction (PCR), small segments of DNA of a defined length are added to prime the site of initiation of DNA synthesis by DNA polymerase. These are called oligonucleotide primers. [☞ Oligonucleotide]

Probe

A relatively small piece of DNA that is used to find another piece of DNA. In nucleic acid hybridization, a DNA probe, labeled radioactively or non-radioactively, seeks out and finds complementary DNA in the target (patient) DNA that is part of the laboratory test. Based on the tag used to label the DNA, different methods are employed to detect that hybridization and answer questions about the patient's DNA that are relevant to the diagnostic issue at hand. The shortest useful probe is about 20 bases long and is known as an oligonucleotide probe. Ligase chain reaction makes use of probes of about this length. Probes can be many hundreds of bases long, and this is the traditional length in laboratory tests like the Southern blot.

RNA molecules may also be used as probes and are termed "riboprobes." [☞ Autoradiogram; Chemiluminescence; Complementary Strands of DNA; DNA Labeling; Hybridization; Southern Blot]

Promoter

Just as prizefighters need the very best promoters (if you're a boxing fan, a particular image has already popped into your mind) to advance their pugilistic careers, so too do genes need promoters. In molecular biology, a promoter is a stretch of bases just upstream (in front of) from the start of a gene. Gene expression begins with transcription of DNA into RNA, and this begins at the so-called transcription initiation site. An enzyme called RNA polymerase, whose job is to synthesize an RNA transcript from a DNA template, binds to the transcription initiation site, also called the promoter. Once bound, RNA polymerase can carry out its task.

Proto-oncogene

Proto-oncogenes are normal genes whose job is to regulate and control cell division and the rate of cellular growth. Proto-oncogenes gone bad, through mutation, chromosomal translocation, etc., lose this important ability, leading to the formation of cancer-causing oncogenes and different kinds of cancer. [☞ Oncogene]

Pseudogene

We have vestiges of genes left in our genomes whose function has been eliminated by evolution but parts of these genes just keep hanging around. These are called pseudogenes. They don't give rise to any functional gene products. Pseudogenes can be a nuisance when you're designing a diagnostic laboratory test searching for a particular gene or gene sequence that you think has relevance to disease; it's possible that a pseudogene sequence can get in the way of a reliable test result.

Purines

A class of nucleotide bases which includes adenine and guanine. [☞ Nucleotide]

Pyrimidines

Another class of nucleotide bases which includes cytosine, thymine, and uracil. [☞ Nucleotide]

Qβ Replicase

No, I didn't make this up just because I needed a "Q" entry. Qβ Replicase is an enzyme. It is an RNA-dependent RNA polymerase, which means that it makes RNA using a parent RNA strand as the template for the daughter strand that is about to be synthesized. The enzyme comes from a bacteriophage called Qβ that naturally contains this enzyme. [☞ Bacteriophage] One company has tried to exploit this enzymatic system for a PCR "wannabe" in the form of a very rapid, very powerful system. [☞ PCR "Wannabes"]

Recombinant DNA (rDNA)

Recombinant DNA (rDNA) is DNA that has been artificially created for purposes of genetic engineering. The goal of the genetic engineering may be industrial-scale production, creation of medical laboratory reagents, or research. [☞ Genetic Engineering; Restriction Endonucleases]

rDNA should not be confused with rRNA: ribosomal RNA, a constituent of ribosomes, which are part of the protein-synthesizing machinery of the cell. The situation gets even more confusing because rDNA is also the abbreviation for that portion of the genome that encodes the sequence of ribosomal RNA.

Replication

I bet you thought that replication was the process of making food and drink in those fancy holes in the wall of the Enterprise on Star Trek. With respect to molecular biology, replication is the process of making more DNA.

Restriction Endonucleases (REs)

Restriction endonucleases, also called restriction enzymes, belong to the general class of enzymes known as nucleases. [☞ Nuclease] Today, there are hundreds of REs available commercially. We now take this for granted. It was only about 20 years ago that REs had to be painstakingly purified in order to proceed with the work of molecular biology. The current abundance and commercialization of REs is an important contributing factor to the existence and growth of the field clinical molecular pathology.

REs are naturally occurring proteins purified from bacteria. Bacterial REs recognize unique stretches of bases, or sites (often, but not always, six base pairs long), in foreign DNA and cleave DNA at or near these sites. Some REs are sensitive to methylation patterns in their recognition sequences. REs are the bacterial immune system. When a bacterium is infected by a bacterial virus (bacteriophage), one way in which that infection can be defeated is for the bacterial RE to recognize the viral DNA as foreign, cut and inactivate it, and thereby prevent the virus from doing its dirty work. We have learned how to use REs for medical diagnostics and genetic engineering.

Recombinant DNA technology has depended on REs because when a piece of DNA is cut by such an enzyme, it leaves an end (sometimes called a "sticky end") on the cut DNA molecule that fits very nicely and can be neatly "sewed" into any one of a number of different cloning vectors, like plasmids or YACs. [☞ Plasmid; Yeast Artificial Chromosome (YAC)] These so-called recombinant DNA molecules can be mass produced.

REs digest DNA specifically, such that unique DNA restriction fragment families are generated by each enzyme. This *in vitro* biochemical reaction generates a range of DNA fragments differing in molecular weight and is the basis for the Southern blot technique that is so important in clinical molecular pathology.

REs are named based on the bacteria from which they are purified. For example the enzyme, *Eco* RI is purified from a particular strain of *Escherichia coli*. Jargon in the laboratory has evolved to the point where we use the names of the enzymes to denote what was done to a particular DNA sample. So we often say that we Bam'ed some DNA when the RE is *Bam* HI, or that we Bgl'd it when the enzyme is *Bgl* II. And yes, there really is an enzyme called *Fok* I.

Restriction Fragment Length Polymorphism (RFLP) Testing

Individuals differ in many ways, including their DNA sequences. One individual may have a restriction endonuclease recognition site at a particular point in his or her DNA while another person may not. [☞ Restriction Endonucleases; Southern Blot] The two sites in these two people are said to be polymorphic, or different, from each other. The difference (polymorphism) is in the length of the restriction fragment generated when those two DNAs are cut with the same restriction endonuclease; hence the name "restriction fragment length polymorphism." RFLPs may be detected by Southern blot analysis or Polymerase Chain Reaction. RFLP analysis has applications in paternity testing, genetic disease, and routine molecular pathology investigation.

Retroviruses

Some viruses have an RNA genome which is converted into a DNA intermediate by an enzyme called reverse transcriptase. [☞ Reverse Transcriptase] That DNA intermediate can become more or less permanently integrated into the host-cell genome; at this part of the viral life cycle the virus is known as a provirus. The virus can leave the proviral stage of its life cycle and synthesize, pirating the machinery of the cell to do its work, new viral proteins and nucleic acids for the creation of viral progeny. The progeny then leave the infected cell and go on to infect new cells.

Retroviruses are RNA tumor viruses. They can cause tumors in the animals they infect, such as monkeys, chickens, rats, and mice. Human T cell Lymphotropic virus, types 1 and 2 (HTLV-1 and HTLV-2) are human retroviruses associated with leukemia and lymphoma. HTLV-3 has had several names but the one most people associate with it is

Human Immunodeficiency Virus (HIV), the agent that causes Acquired Immunodeficiency Syndrome (AIDS).

Reverse Transcriptase (RT)

Reverse transcriptase is a nucleic acid polymerase; that is, it's an enzyme (protein) whose job is to make nucleic acid using another nucleic acid as a template for that synthesis. RT is an RNA-dependent DNA polymerase; it makes DNA using an RNA template to direct the synthesis of that DNA.

RT is the exception I wrote about in the entry "Gene Expression." For years it was believed that the flow of genetic information proceeded in only one direction: DNA → RNA → Protein. In fact, this belief was so strong that it was considered the central dogma of molecular biology. And we all know what happens when we become dogmatic: We're usually proved wrong. That's exactly what happened here.

Research findings were published independently in 1970 by three scientists, two of whom would go on to win the Nobel Prize for this work. David Baltimore at the Massachusetts Institute of Technology and Howard Temin at the McArdle Laboratory for Cancer Research at the University of Wisconsin in Madison (working with his colleague Satoshi Mizutani) did their studies on a class of viruses known as RNA tumor viruses. These are viruses which have only RNA in their genomes (when examined outside the cells they infect), and which under proper conditions cause tumors in the animals they infect. Temin (and Mizutani) and Baltimore went on to show that these viruses contain an enzyme which has the ability to direct the synthesis of DNA using the original viral RNA as a template. The enzyme, called RNA-dependent DNA polymerase, was dubbed "reverse transcriptase" because it shattered the then-current dogma by showing that transcription could happen in a "reverse" way. Transcription became a more generalized term to indicate the formation of intermediary nucleic acid (usually RNA, but it was now shown that DNA could be that intermediary) which went on to direct the rest of the viral life cycle within the infected cell.

The flow of genetic information in this class of viruses, which went on to become known as "Retroviruses," is RNA → DNA → RNA → Protein. Infecting viral RNA, through the action of virally encoded reverse transcriptase, is transcribed into a DNA intermediary which goes on to direct the synthesis of viral RNA (for progeny virus); ultimately, viral proteins for new viral progeny are also made. The virus can arrest in its life cycle at the DNA stage. The DNA becomes integrated into the host cell genome, and it is known as a provirus at this point. [☞ also Gene Expression; Retroviruses; Virus]

RFLP

[☞ Restriction Fragment Length Polymorphism (RFLP) Testing]

Ribonucleic Acid (RNA)

Weak sister of DNA and the intermediary for the flow of genetic information. (Don't women usually wind up doing most of the work anyway?) DNA–admittedly a prima donna–just sits there getting all the glory, press, and notoriety while RNA and proteins within the cell work to replicate it, proofread it, and allow it to express itself. [☞ also DNA; Messenger RNA (mRNA); Nucleotide; Nucleic Acid; Ribosomal RNA (rRNA)]

Riboprobe

[☞ Probe]

Ribosomal RNA (rRNA)

Ribosomal RNA is a constituent of ribosomes. Ribosomes are a large part of the protein-synthesizing machinery of the cell.

RNA

[☞ Ribonucleic Acid]

RNase

[☞ Nuclease]

RT-PCR

Think of it as RNA-PCR. [☞ PCR (Polymerase Chain Reaction)]

Safety

When handling patient specimens in the laboratory, we are always concerned about safety. In fact, we treat every specimen as if it actually were contaminated with human pathogens, like HIV-1 or Hepatitis C virus. For this reason, human tissues, all of which *may* contain unknown infectious agents, are handled using "universal precautions." These are guidelines which, when followed properly, minimize health risks to laboratory workers associated with the handling of human specimens. The guidelines are published in the following document:

CDC/NIH Biosafety in Microbiological and
Biomedical Laboratories, 3rd edition.
U.S. Government Printing Office,
Washington, D.C. 20402
May 1993; $6.75
HHS Publication No. (CDC) 93-8395
U.S. GPO Stock # 017-040-00523-7

More information on this topic is available on the Internet at:

http://www.cdc.gov/ncidod/diseases/hip/prvthivt.htm
http://www.cdc.gov/ncidod/diseases/hip/universa.htm

Salsa

[☞ DNA Chips]

Self-Sustained Sequence Replication (SSSR)

SSSR (also known as 3SR) is an *in vitro* nucleic acid amplification technique. It is a PCR "wannabe." [☞ PCR "Wannabes"]

Semiconservative

When you're young and your mind is full of utopian ideas it's easy to have a liberal point of view on things. As you get older and accumulate more, conservative thinking starts to creep into your psyche. I'm in that awkward transitional period–you could say I'm semiconservative. DNA is also semiconservative, at least with respect to its replication. During DNA replication, one strand of the double helix serves as the template for the synthesis of a new daughter strand. After replication is completed, there is one "old" strand of DNA that served as the template and the complementary, new daughter strand: one old and one new. That's the thought behind terming DNA replication semiconservative.

Slot Blot

[☞ Southern Blot]

Southern Blot (Dot Blot, Slot Blot)

The technique known as the Southern blot was developed by Dr. E. Southern in the mid 1970s in England. It is a method of looking for a needle in a haystack, or more correctly, a single piece of hay within the haystack. Purified DNA is subjected to fragmentation with restriction endonucleases and electrophoresis to separate those fragments. [☞ Electrophoresis; Restriction Endonucleases] The gene or DNA sequence of interest is still buried within that total genome's worth of DNA. The next step is to transfer the DNA in the gel used for electrophoresis to a more solid support. [☞ Agarose] Early investigators used nitrocellulose paper; the field now uses nylon membranes for the most part. Either way, DNA is transferred to a solid support which is little more than a very tough piece of special paper. This is accomplished by vacuum (actually aspirating the DNA out of the gel and onto the paper), positive pressure, or capillary action (where salt water moves up from a reservoir through the DNA-containing gel and carries or transfers the DNA to a piece of nylon paper on top of the gel–the DNA binds to the paper). After the transfer is complete, the association between DNA and paper is tenuous but is made permanent by baking the Southern blot in an oven for an hour or two (about 80° C). Once permanently attached, the DNA on the blot may be hybridized with a DNA probe to actually locate the DNA sequence of interest: the proverbial needle in the haystack. [☞ Autoradiogram; Probe]

Southern blot based testing is most often used in the clinical molecular pathology laboratory for B and T cell gene rearrangement analysis, *bcr* gene rearrangement analysis, and fragile X syndrome analysis. There are also other applications.

Dot and slot blots are variations on the theme of the Southern blot. They are the same except no restriction endonuclease digestion of DNA and subsequent electrophoresis is done. Dot and slot blots, therefore, are used more for simple "yes or no" answers to the question: "Is a particular DNA sequence (for example, a mutation or bacterial DNA) present in this sample?"

Splice Site Mutations

[☞ Gene Splicing; also Codon; Genetic Code; Genotype; Mutation; Open Reading Frame]

SSSR

[☞ Self-Sustained Sequence Replication]

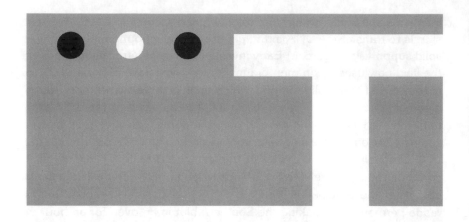

Telomere/Telomerase

Telomeres are located at the ends of chromosomes. In humans, the telomere sequence is TTAGGG, repeated hundreds and up to a thousand times. This telomeric sequence is synthesized by a cellular enzyme called telomerase and is necessary to counter the normal shrinkage at the ends of chromosomes that occurs after every round of DNA replication. Telomerase is expressed in our eggs and sperm, so these telomeres do not shorten throughout life and these cells are able to divide throughout life. We lose 15 to 40 nucleotides per year off the ends of the chromosomes in our skin and blood cells, which do not express telomerase.

Many tumor cells have been found to express telomerase, and it has been suggested that in this way tumor cells escape the normal cellular mechanism that nature has installed to regulate the life span of normal cells like skin and blood cells. In other words, the abnormal expression of telomerase in cancer cells may be the way these cells become immortal. It remains to be seen, through basic research, whether attacking telomerase activity in cancer cells will stop the growth of a particular tumor.

Template

When DNA or RNA is replicated, either naturally *in vivo,* or artificially in the laboratory *in vitro,* new nucleic acid is being made, by definition. That synthesis is dependent upon the action of enzymes. The polymerase involved in the creation of new DNA or RNA is dependent upon a master copy to direct the synthesis of the new strand of nucleic acid. That master copy is known as the template. [☞ Complementary Strands of DNA; Polymerase]

Tm

T_m is the melting temperature of a DNA duplex, defined as the temperature at which 50% of the double-stranded DNA molecules in solution are dissociated from each other and 50% are associated with each other. G-C base pairs in DNA are more stable than A-T base pairs because G-C pairs have 3 hydrogen bonds holding them together and A-T pairs have only 2. Therefore, the higher the G-C content of a particular piece of DNA, the more thermal energy required to dissociate the DNA strands, and the higher the T_m.

Thermal Cycler

A thermal cycler is a microprocessor controlled water bath that rapidly changes the temperatures among those needed to accomplish the Polymerase Chain Reaction (PCR). A typical PCR may have to cycle among 94° C, 55° C, and 72° C 30 or 40 times. Thermal cyclers accomplish this in an automated fashion. [☞ PCR]

Thermus aquaticus

An important technical advance for Polymerase Chain Reaction (PCR) came with the realization that there are bacteria that normally carry on the business of life in hot springs (like the ones in Yellowstone National Park). The bacterium, *Thermus aquaticus,* lives in such springs at temperatures of 75° C and the DNA polymerase purified from this bacterium functions at temperatures over 90° C. This DNA polymerase, named *Taq polymerase* after the bacteria from which it is purified (*Thermus aquaticus*), is the workhorse of PCR and has been a significant factor in the wide use of PCR in the clinical molecular pathology laboratory. [☞ PCR]

Tissue-Specific Gene Expression

All the cells in our bodies, except mature red blood cells and gametes, [☞ Allele] contain the full measure of DNA that we have. Mature red blood cells contain no DNA. Gametes (sperm and eggs) contain half the DNA found in a somatic cell, like a stomach, nerve, or liver cell. By "full measure" I mean that even though a stomach cell is specialized for the tasks that it does, it still has all the DNA that is involved in hair color, antibody production, vision, and everything else that our bodies do that depends on the expression of proteins encoded by DNA.

What makes a stomach cell a stomach cell and not a liver cell or nerve cell is that, due to complex biological, endocrinological, biochemical, and other processes, tissue-specific gene expression occurs. A cell committed to being a stomach cell and finding itself in the biological environment of the stomach expresses only those genes in its full complement of DNA that are necessary for it to function as a stomach cell. All the other DNA contained in the cell remains silent and unexpressed, so that the cell doesn't start doing the job of a liver cell, a nerve cell, or some other cell. Tissue-specific gene expression is what allows the different cells in our bodies to specialize and form specialized tissue, organs, and organ systems. Without it, we would experience total biological chaos.

Transcription

The synthesis of RNA and mRNA from a DNA template is the process known as transcription. [☞ Messenger RNA (mRNA); Ribonucleic Acid (RNA)]

Transfer RNA (tRNA)

An RNA molecule that transports amino acids to the ribosome for protein synthesis. [☞ Anticodon]

Transgenic Animal

A transgenic animal is one that has had a foreign DNA sequence, i.e., a gene, introduced into it early in its development, even as early as when the animal was a single cell. The process is done by holding the cell in place and microinjecting DNA into it with the aid of a microscope. The research opportunities afforded by transgenic animals have been bountiful, allowing us to learn more about gene expression, gene regulation, cancer formation, and more. The practical repercussions may be very significant as commercial and research endeavors move towards developing medically useful transgenic animals. Examples include transgenic goats whose blood contains human proteins and is therefore suitable for human blood transfusion; and transgenic pigs that have human surface proteins in their hearts, which are not rejected as "foreign" when transplanted into a human. The medical implications are exciting. At the same time, as a society we need to deal with the issue of the ethical treatment of animals. This is one of the difficult questions that scientific progress is forcing us to consider.

Translation

The synthesis of proteins from a mRNA molecule using ribosomes and the rest of the cell's protein synthesizing machinery is called translation.

Triplet

[☞ Genetic Code]

Tumor Suppressor Genes

Tumor suppressor genes have been implicated in the pathogenesis of different cancers that occur as rarely as retinoblastoma (an eye tumor) and as frequently as colon cancer. Normally there are 2 alleles of a gene on chromosome 13 called *RB*. Loss of one copy through mutation does not affect the individual or lead to cancer, but loss of the second allele through mutation leads to deregulated cell growth and retinoblastoma. The normal function of this gene is to suppress the growth of this kind of tumor. When that gene's function is lost due to mutation, the normal check on tumor development is lost, much like losing the brakes on a car.

The p53 tumor suppressor gene, when mutated, appears to be involved in many different kinds of cancer, including breast cancer, and may very well be an important object of clinical investigation in the years to come as we learn more about this gene and its mutations. A test for p53 mutations may become an important step in diagnosing cancer risk so that early intervention and possible avoidance may occur. The p53 gene is so named because the gene product is a protein that is 53,000 daltons in mass.

Virus

Those who are inclined to ponder such things generally consider viruses to be the simplest and most basic form of life, although the prions of Mad Cow Disease fame might argue the point. Viruses reproduce (one of the most basic definitions of life) by commandeering the cellular machinery of the cells they infect. Those cells could be animal cells (eukaryotic cells: cells with a distinct nucleus and several other characteristic features), and the viruses that pirate them are called animal viruses. The viruses that infect bacterial cells (prokaryotic cells: cells which are much more primitive than eukaryotic cells) are called bacterial viruses or bacteriophage ("phage" is another word for virus). Viruses are obligate intracellular parasites; they cannot reproduce without first entering a host cell.

Viruses have another unifying feature in addition to their life cycles (which vary considerably but all depend on parasitism): Viruses have one kind of genetic material–DNA or RNA but not both–and that nucleic acid can differ: there are single-stranded RNA viruses, such as Human Immunodeficiency Virus (HIV), Hepatitis C virus, and Rhinovirus; double-stranded RNA viruses, such as Reovirus and Colorado tick fever virus; single-stranded DNA viruses, such as Parvovirus; and double-stranded DNA viruses, such as Herpes Simplex Virus 1 and 2 and Poxvirus.

The viral genome is surrounded by a protein coating or shell called the capsid. The genome plus the capsid is called the nucleocapsid. There are viruses that consist only of naked nucleocapsids while others have an enveloping layer surrounding them that is composed of lipids (fats) and glycoproteins. An intact viral structure that has the ability to infect is commonly referred to as a virion.

There are several ways that different viruses complete their life cycles within their host cells (always keeping in mind that the host cell would rather be rude and kick the "guest" out). Some viruses infect a cell and go about the business of making progeny virus. They do this by directing the cell's transcription and translation machinery to recognize the genetic material of the virus (RNA or DNA) and do its bidding. The bidding of that genetic material is actually the viral proteins encoded in the virus's RNA or DNA, and the infection proceeds such that more viral proteins and nucleic acids are made. Ultimately, these viral subunits made by the cell, under the direction of the virus, accumulate to form progeny virus.

The mechanism of release from the cell can be violent and catastrophic for the cell or it can be more protracted. Some viruses accumulate to such large numbers that the cell ultimately swells and bursts (the technical term for this is "lyse"), releasing thousands to millions of new, infectious virions which then go on to repeat the cycle by infecting other cells. Some viruses, principally those that are naturally enveloped (a fatty, more or less circular covering around the outside of the virus) accumulate progeny virus within the infected cell more slowly and actually "bud" in relatively smaller numbers from the outer cell membrane.

Some viruses like their new homes (infect), unpack (shed their fatty, protein coats), move in (insert their genetic material into the genome of the host cell), freeload (direct the cell to perform functions necessary to maintain the viral genome in the host genome without destroying the host cell), and ultimately are not embarrassed about deserting a sinking ship like any rat. (A variety of things can cause an embedded virus, called a provirus, to "decide" it's time to leave.) When the virus leaves the so-called latent phase and enters a more active phase, new viral particles are made and leave the cell, which they may or may not kill in the process.

Because viruses are so small and simple, decades of research into how they work have been a rich source of information, not only about virology, but also about basic genetics and the biochemistry of DNA and RNA replication, protein synthesis, viral pathogenesis, and much more.

Watson, James D.

Francis H. C. Crick and James D. Watson (with a little help from their professional colleagues) deduced the double-helical nature of DNA, realized how that structure lent itself to replication of the molecule, shared the 1962 Nobel Prize for their work (along with Maurice Wilkins), went on to publish many more scientific manuscripts, write books, give talks, become faculty, and head scientific research institutes. For their work, scientists affectionately call one strand of double-stranded DNA the "Watson strand" and the other strand the "Crick strand."

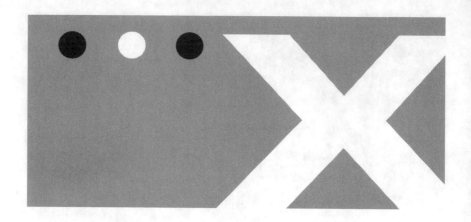

X Chromosome

I have a five-year-old son, Joshua (he has one X chromosome in his cells, like all normal males) and a three-year-old daughter, Haley (she has two X chromosomes in her cells, like all normal females). So I am very familiar with children's books that attempt to teach the alphabet. Haven't you ever noticed that they always struggle for an "X" entry; it's usually xylophone or X-ray and beyond those two it's always a reach. I got lucky here because X chromosome is an important entry for this book.

In normal females two X chromosomes are present per cell. But females don't express twice as many of the proteins encoded by genes on the X chromosome as males, who have only one X chromosome per cell. In 1961, Dr. Mary Lyon hypothesized that one of the X chromosomes in the female cell is inactivated or shut down, precisely to avoid this problem. This X chromosome inactivation is known as Lyonization. [☞ Y Chromosome]

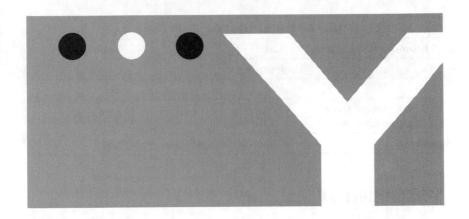

Y Chromosome

Male gender (sex) is determined by the appropriate expression of key genes, the interaction of genes and gene products with different proteins, including hormones, and a key gene present on the Y chromosome in mammals (including man) called the *SRY* gene, for sex-related Y. Sex differentiation *in utero* is a complicated, intricate weave of gene expression and the influence of hormones. But even a point mutation in the *SRY* gene can cause an XY individual, who would normally have a male phenotype, to have an incomplete female phenotype. [☞ Mutation; Phenotype] Some have suggested that certain characteristics are associated with *SRY*, including but not limited to: refusal to ask for directions when lost, channel-surfing, and an inability to offer much sympathy.

Testicular feminization is an abnormal condition in individuals who are XY and should have a male phenotype, but who cannot use male hormones properly inside the relevant cells that depend on these hormones. These individuals are outwardly female and many are quite striking. My freshman biology professor used to say, "If it's too good to be true, suspect testicular feminization" (thank you, Dr. Bromley). [☞ also X Chromosome]

HUMAN CHROMOSOME PATTERNS		
Chromosomal Pattern	Label	Phenotype
XX	Female	Female
XY	Male	Male
XO (1 X chromosome; no Y chromosome)	Turner Syndrome	Female
XXY	Kleinfelter's Symdrome	Male

Yeast Artificial Chromosome (YAC)

You know those large, woolly animals at the zoo that look something like a cross between a woolly mammoth and a cow? Well, those are yaks; YACs are cloning vectors used in the DNA laboratory. A cloning vector is a piece of DNA that has the capability of being replicated or duplicated *en masse*. Scientists who want to use or study a specific piece of DNA can generate lots and lots of copies of it by inserting it into a cloning vector, such as a YAC. YACs replicate to high copy numbers inside yeast cells. A plasmid, another piece of DNA which is popular for use as a cloning vector, replicates to high copy numbers in bacteria. Both yeast and bacteria are life forms that we can grow and multiply in the laboratory. As they increase in number, so too do the cloning vectors, with the piece of DNA of interest inside them. [☞ also Genetic Engineering; Plasmid; Recombinant DNA]

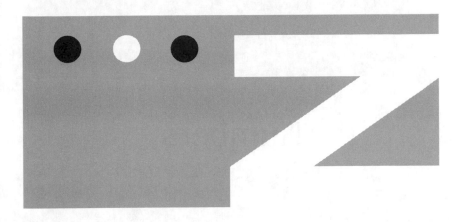

Z-DNA

I needed a "Z" entry. Naturally occurring DNA, called B-DNA, has a right-hand turn (like an ordinary wood or metal screw) to the double helix. Another form of DNA has been observed that has a left-hand turn to its helical structure. There's a lot of physical chemistry involved that causes the nucleotides within Z-DNA to course through the helix in a sort of zig-zag manner, which is why it was termed Z-DNA. [☞ also A-DNA; B-DNA]

Numbers

21mer

21 happens to be a very good length probe and primer (and is also a very good number at the Black Jack table). With respect to DNA, "21mer" refers to a stretch of synthetically prepared DNA that happens to be 21 bases in length. There's no reason you couldn't have a "16mer" or a "37mer," etc. Probes and primers are described above. [☞ Probe; Primers] It turns out that 21mers are a popular length probe or primer because–for physical and chemical reasons–they serve as nicely stable and specific probes for target sequences of DNA that are perfectly complementary to them. [☞ also Complementary Strands of DNA]

46

The number of chromosomes in a cell. *Homo sapiens* (humans) have 22 pairs of autosomes (non-sex chromosomes) in each cell (except red blood cells) and 1 pair of sex chromosomes for a total of 23 pairs (46 in all). Females have 1 pair of sex chromosomes: two X chromosomes; males also have 1 pair of sex chromosomes: one X and one Y chromosome.

1953

The year James D. Watson and Francis H. C. Crick published their elucidation of the structure of DNA. The article was entitled: "Molecular Structure of Nucleic Acids: A Structure for Deoxyribose Nucleic Acid" and was published in the British journal, *Nature*, volume 171, page 737, April 25, 1953. The paper was followed quickly by another by Watson and Crick in which they more fully described the replication process for DNA: "Genetic Implications of the Structure of Deoxyribonucleic Acid" published in *Nature*, volume 171, pages 964–967, May 30, 1953.

1958

The year my genetic material was expressed (actually, since I was born in May 1958, gene expression began in 1957, but that kind of stuff is way too Freudian for me to think about comfortably). Dwight D. Eisenhower was President of the United States; Joseph I. Routh, PhD, was President of the American Association for Clinical Chemistry until the Annual Meeting in mid-year when Oliver H. Gaebler, PhD, assumed the office, and the New York Yankees beat the Milwaukee Braves in the World Series (Mickey Mantle hit 2 home runs in the series; Henry Aaron didn't hit any). There was no Super Bowl yet, and my brothers were 12 and 9 years old. My wife's genetic material had not been united *in utero* yet; that wouldn't occur for another two years.

1990

The year my son's genetic material was expressed.

1993

The year my daughter's genetic material was expressed.

60,000 (6×10^4)

The estimated number of human structural genes.

3,000,000,000 (3×10^9)

The number of nucleotide base pairs in a human sperm or egg cell. [☞ Complementary Strands of DNA] Non-sex cells (somatic cells) like stomach, nerve and muscle cells have twice as much DNA. Here are some fairly useless arithmetic facts:

If you multiply the number of base pairs in a somatic cell (6×10^9) by the length of the DNA purified from a *single* cell and string out that DNA in a straight line (3.4×10^{-10} meters per base pair) the product is about 2.04 meters of DNA per cell. If you multiply 2.04 meters of DNA in one cell by the number of cells in a mature adult human (about 3.5×10^{13}) the product is 7.14×10^{13} meters. (714 is also the number of home runs Babe Ruth hit in his major league career.) The distance from the Earth to the Sun (one way) is about 93,000,000 miles. One mile is about 1625 meters, so the distance to the Sun is about 1.5×10^{11} meters. If you divide 7.14×10^{13} by 1.5×10^{11}, you find that the number of meters of DNA in one person could be strung back and forth between the Sun and the Earth 476 times. That's a lot of DNA.

98

Appendix

Professional societies and organizations whose membership is involved in medical research and in the implementation, certification, education, training, and proper usage of DNA technology in the clinical laboratory, or otherwise involved with "DNA." The following is *not* a comprehensive list.

AABB	American Association of Blood Banks	Bethesda, MD
AACC	American Association for Clinical Chemistry	Washington,DC
ACMG	American College of Medical Genetics	Bethesda, MD
AMP	Association for Molecular Pathology	Bethesda, MD
ASCP	American Society of Clinical Pathologists	Chicago, IL
ASHG	American Society of Human Genetics	Bethesda, MD
ATCC	American Type Culture Collection	Rockville, MD
CAP	College of American Pathologists	Northfield, IL

DOE	Department of Energy Washington, DC
ELSI	Ethical, Legal and Social Implications (of the Human Genome Project)* Washington, DC
FBI	Federal Bureau of Investigation Washington, DC, and Quantico, VA
HUGO	Human Genome Organization (international)
MRC	Medical Research Council (United Kingdom)
NCA	National Certification Agency Lenexa, KS
NCI	National Cancer Institute Bethesda, MD
NIAID	National Institute of Allergies and Infectious Disease Bethesda, MD
NIDDK	National Institute of Diabetes and Digestive and Kidney Diseases, Bethesda, MD
NIH	National Institutes of Health Bethesda, MD
OSHA	Occupational Safety and Health Administration Washington, DC
USCAP	United States and Canadian Academy of Pathology Augusta, GA

A list of available training may be viewed on the Internet at:

http://www.ornl.gov/TechResources/Human_Genome/
publicat/hgn/v7n5/trcal.html

*James Watson, Nobel Laureate and co-discoverer of the double-helical structure of DNA, with vision and thought, earmarked a significant portion of Human Genome Project funds for ELSI when he was head of the Project.

Further Reading

If this book has whetted your appetite for more serious and intensive treatment of DNA and related subjects, a wealth of further information is available. For Internet browsers, just type in key words like "DNA," "Genetics," or "Molecular Pathology" and you'll be amazed at how much information pops up.

For those of you who would like to do further reading, following are some books that I use a lot and find helpful and informative:

1. Andrews LB, Fullarton JE, Holtzman NA, Motulsky AG, eds. *Assessing genetic risks: Implications for health and social policy.* Washington, DC: National Academy Press, 1994.

2. Bernstam VA. *Handbook of gene level diagnostics in clinical practice.* Boca Raton, FL: CRC Press, 1992.

3. Culver KW. *Gene therapy: A handbook for physicians.* New York, NY: Mary Ann Liebert, Inc., 1994.

4. Farkas DH, ed. *Molecular biology and pathology: A guidebook for quality control.* San Diego: Academic Press, 1993.

5. Friedman T, ed. *Molecular genetic medicine.* Academic Press, San Diego, CA. 1994.

6. Heim RA, Silverman LM, eds. *Molecular pathology: Approaches to diagnosing human disease in the clinical laboratory.* Carolina Academic Press, Durham, NC. 1994.

7. Herrington CS, McGee JO'D, eds. *Diagnostic molecular pathology.* New York: Oxford University Press, 1992.

8. Innis MA, Gelfand DH, Sninsky JJ, White TJ, eds. *PCR protocols: A guide to methods and applications.* San Diego: Academic Press, 1990.

9. Kirby LT. *DNA fingerprinting: An introduction.* New York: Stockton Press, 1990.

10. Lee TF. *The Human Genome Project: Cracking the genetic code of life.* New York: Plenum Press, 1991.

11. Mullis, KB, Ferre, F, Gibbs, RA, eds. *The polymerase chain reaction.* Boston, MA: Birkhäuser Press, 1994.

12. Passarge E. *Color atlas of genetics.* New York: Thieme Medical Publishers, 1995.

13. Persing DH, Smith TF, Tenover FC, White TJ, eds. *Diagnostic molecular microbiology: Principles and applications.* Washington DC: American Society for Microbiology, 1993.

14. Watson JD, Gilman M, Witkowski J, Zoller M. *Recombinant DNA,* 2nd ed. [Scientific American Books.] New York: WH Freeman, 1992.

15. Wiedbrauk DL, Farkas DH, eds. *Molecular methods for virus detection.* San Diego: Academic Press, 1995.

16. Williams RC. *Molecular biology in clinical medicine.* New York: Elsevier, 1991.

About the Author

Photo by Peter Roberts

Daniel H. Farkas was born and raised in Brooklyn, New York. He received his B.S. degree in Microbiology and Public Health from Michigan State University in East Lansing. He received a Ph.D. in Cellular and Molecular Biology from the State University of New York at Buffalo, Roswell Park Cancer Institute Graduate Division, in 1987. After two years as a Postdoctoral Fellow in St. Louis, Dr. Farkas established and ran the DNA Diagnostics Laboratory within the Department of Pathology at Saint Barnabas Medical Center in Livingston, New Jersey. In 1991 he joined the medical staff at William Beaumont Hospital in Royal Oak, Michigan, where he holds his current position as co-director of the Molecular Probe Laboratory.

Dr. Farkas is active in many professional organizations including the American Association for Clinical Chemistry (AACC), the Association for Molecular Pathology (AMP), the National Certification Agency, and the American Society of Clinical Pathologists. He has served on two College of American Pathologists (CAP) committees and currently sits on the CAP Molecular Pathology Resource Committee as AACC liaison. Dr. Farkas was chairperson of the AACC's Molecular Pathology Division in 1994, is currently the editor of the AMP Newsletter and serves on AMP's Clinical Practice Committee. In 1995, he was Program Director for AACC's premier symposium on DNA technology, the San Diego Conference on Nucleic Acids.

Since 1992, Dr. Farkas has organized a yearly symposium on "DNA Technology in the Clinical Laboratory" at William Beaumont Hospital. He is on the clinical faculty of the School of Medical Technology within the Department of Clinical Pathology at William Beaumont Hospital and is an associate adjunct professor of Medical Technology at his alma mater, Michigan State University.

Dr. Farkas serves on the Editorial Board of *Diagnostic Molecular Pathology*, *Molecular Diagnosis*, and *Clinical Laboratory News*. He has lectured internationally on the topic of DNA diagnostics, has published over thirty papers in the field, and has edited two volumes of *Clinics in Laboratory Medicine* on the subject of DNA technology. This is his third book on DNA and clinical molecular biology.

Dr. Farkas lives in Rochester Hills, Michigan, with his wife, Becky, his son, Joshua, age 5, and his daughter, Haley, age 3. He still roots for the New York Mets, Jets, Rangers, Knicks, and the Michigan State Spartans.

Index

carcinogenesis. *See* cancer
cDNA, 12–13, 16–17, 67
centric fusion, 13
chemiluminescence, 13. *See also*
  autoradiograph
Chlamydia trachomatis, 70
chromosomal translocation, 13–14,
  35. *See also* PCR; Southern Blot
chromosome
    number of, 2–3, 14, 96
    structure, 13, 84
    translocation, 13–14, 35
    X, 92
    Y, 93
cistron, 15. *See also* gene
CLIA '88, 15
Clinical Laboratory Improvement
  Amendments of 1988, 15
clone, 15–16, 35, 94
codon, 16. *See also* anticodon;
  genetic code
complementary strands of DNA,
  12–13, 16–17, 18, 51–52.
  See also denature
Crick, Francis H. C., 17, 91, 97
cystic fibrosis, 41, 55
cytosine, 16–17

ddNTP, 22
denature, 18
deoxyribonucleic acid. *See* DNA
deoxyribonucleotide, 5
dideoxy sequencing, 22
diploid, 2–3
DNA, 18–19. *See also* nucleic acid;
  nucleotide; RNA
    A-DNA, 1–2
    amount in human body, 8–9, 98
    annealing, 65–66
    bank, 19–20
    B-DNA, 7

DNA *(Continued)*
    bDNA, 7, 69
    cDNA, 12–13, 16–17, 67
    in cell types, 86
    chip, 20
    degradation, 57
    denaturing, 18, 65–66
    discovery of, 51
    electrical conductivity, 24
    extraction/purification, 21
    fingerprinting, 64–65
    junk, 45
    labeling, 21, 46, 56, 61
    lambda, 48
    melting temperature, 85
    minisatellite, 63–64
    mitochondrial, 52
    mutation, 53–55
    polymerase, 71–72, 78–79
    recombinant, 39–40, 75, 76–77
    replication, 27–28, 47, 71–72,
      75, 82, 84–85
    restriction, 76–77
    sequencing, 22–23, 24–25, 43
    structure, 1–2, 23, 50, 51, 95
      antiparallel, 5
      complementary strands, 16–17
      histone, 42
      nucleotide, 9, 37–39, 59
    transcription, 72–73
    virus, 89–90
    Z-DNA, 95
DNA bank, 19–20. *See also* pater-
  nity/profiling/identity/forensic test-
  ing by DNA
DNA chip, 20. *See also* oligonucleo-
  tide
DNA extraction/purification, 21
DNA fingerprinting, 64–65
DNA labeling, 21, 46, 56, 61
DNA replication, 27–28, 47, 71–72,
    75, 82, 84–85

DNase. *See* nuclease
dot blot. *See* Southern blot
duplex, 23

electric gene, 24
electrophoresis, 2, 8, 24–25, 48
enhancer, 26
enzyme, 1, 26
ethidium bromide, 8, 26–27
exon, 27, 36. *See also* gene expression
expression. *See* gene expression
extension, 27–28. *See also* PCR

familial hypercholesteremia, 54
forensic DNA testing, 63–65, 77
fragile X syndrome, 29–30
frameshift, 38, 54

GC-rich, 31. *See also* nucleotide
GenBank, 31
gender, 93, 94
gene, 32, 46
   allele, 2–3, 40–41
   electric, 24
   expression, 32–33, 72–73, 86
   genetic code, 37–39, 59, 62
   genotype, 40–41
   locus, 49
   number of, 98
   product, 33–34
   pseudogene, 73
   rearrangement, 13–14, 34–35
   splicing, 36
   therapy, 36–37
gene expression, 26, 32–33, 72–73
   tissue-specific, 86
genetic code, 37–39, 59, 62. *See also* gene expression

genetic counseling, 39
genetic engineering, 39–40
genetic testing, 63–65
genome, 40, 43
genotype, 40–41, 70. *See also* allele
guanine, 16–17

hairpins, 42
HGP, 43
histone, 42
Human Genome Project, 43
Human Immunodeficiency Virus, 77–78
Human T cell Lymphotropic virus, 77
hybridization, 3–4, 43, 72. *See also* complementary strands of DNA; probe

identity testing, 63–65, 67
immune response, 34–35
Internet. *See* World Wide Web
intron, 36, 44, 45. *See also* exon
in utero, 44
in vitro, 44
in vivo, 44

junk DNA, 45

kb, 46
kinase, 46
Kleinfelter's Syndrome, 94

labeling. *See* autoradiograph; chemiluminescence; DNA labeling; kinase
laboratory safety, 81
lagging strand, 47

lambda DNA, 48
LCR. *See* ligase chain reaction
leading strand. *See* lagging strand
ligase chain reaction, 48–49, 69, 70, 72
locus, 49
Lyon, Mary, 92
Lyonization, 92

major groove, 50
Mendel, Gregor, 50
Miescher, Frederick, 51
minisatellite repeats, 64
minor groove, 51
mismatch, 51–52
mitochondria, 52
mitosis, 13
molecular biology, 52
molecular pathology, 53
mRNA, 4, 51
mtDNA, 52
Mullis, Kary, 67, 68
mutation, 36, 38, 53–55

NASBA. *See* nucleic acid sequence-based amplification; PCR
National Alliance of Breast Cancer Organizations, 11
National Cancer Institute's Cancer Information Service, 11
National Fragile X Foundation, 30
National Society of Genetics Counselors, 39
nick translation, 56, 61
northern blot, 56–57. *See also* Southern blot
nuclease, 57, 76–77
nucleic acid, 57–58. *See also* DNA; RNA

nucleic acid sequence-based amplification, 58, 69
nucleocapsid, 89–90
nucleoside. *See* nucleotide
nucleosome, 42, 59
nucleotide, 9, 31, 59, 98
numbers, 96–98

OJ, 60
oligonucleotide, 20, 42, 60
   priming, 61, 71–72
oligonucleotide array. *See* DNA chip
oligos. *See* oligonucleotide
oncogene, 61–62
open reading frame, 62. *See also* genetic code
ORF, 62

p53, 88
paternity/profiling/identity/forensic testing by DNA, 63–65, 77
PCR, 65–68. *See also* primer
   Chlamydia trachomatis detection, 70
   oligonucleotide, 60
   polymorphism identification, 64
   in a pouch, 68–69
   RT-PCR, 67
   Taq polymerase, 85–86
   wannabes, 69
PCR in a pouch, 68–69
PCR wannabes, 69
phenotype, 40–41, 70
plasmid, 70, 94. *See also* clone; genetic engineering
polymerase, 71–72, 78–79
polymerase chain reaction. *See* PCR
polymorphism, 63–64
primer, 42, 61, 66–67, 71–72, 96

108

telomere, 84
Temin, Howard, 78
template, 84–85
testicular feminization, 93
beta-thalassemia, 54
thermal cycler, 67–68, 85
Thermus aquaticus, 85–86
thymine, 16–17
tissue-specific gene expression, 86.
  See also gene expression
T_m, 85
transcription, 78–79, 86. *See also*
  RNA
Transcription Mediated Amplification, 69
transfer RNA, 4, 86
transgenic animal, 87
translation, 87
trinucleotide repeat amplification, 30
triplet. *See* genetic code
triploid, 3
tRNA, 4
Tth polymerase, 13, 67
tumor suppressor gene, 87–88
Turner Syndrome, 94
21mer, 96

uracil, 17

Variable Number Tandem Repeats,
  64
vector, 15, 19
virion, 90
virus, 8, 78–79, 89–90
VNTR, 64

Watson, James D., 17, 91, 97
World Wide Web, 10–11, 31, 81, 100

X chromosome, 92

Y chromosome, 93, 94
YAC. *See* Yeast Artificial Chromosome
yeast artificial chromosome, 94. *See
  also* genetic engineering; plasmid;
  recombinant DNA
Y-ME Hotline, 11

Z-DNA, 95